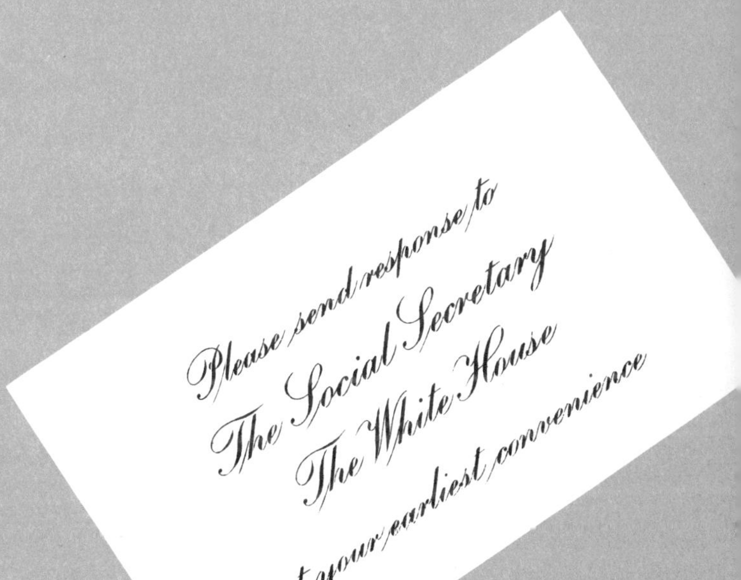
Please send response to
The Social Secretary
The White House
at your earliest convenience

ENTERTAINING IN THE WHITE HOUSE

ENTERTAINING

Published by **Acropolis Books**
Washington, D.C.
AMERICANA *by* ACROPOLIS

IN THE WHITE HOUSE

by MARIE SMITH

Painting on Title Page of John Quincy Adams dancing lively Virginia Reel by National Geographic Artist Robert C. Magis. Copyright by White House Historical Association.

ACROPOLIS BOOKS
Colortone Building, 2400 17th St., N.W.
Washington, D.C. 20009

FIRST EDITION
Printed in the United States of America by
Colortone Creative Printing, Washington, D. C, 20009

Type set in Baskerville
by Colortone Typographic Division, Inc.

Design by Design and Art Studio 2400, Inc.

Library of Congress Catalog Number 67-23364

Contents

Introduction

The most brilliant, beautiful and exciting social events in Washington are the state dinners at the White House. Because they are official functions and represent the hospitality of a prosperous, hospitable nation, every effort is expended to make them as glittering as possible.

Equally as exciting and beautiful in many cases are the other social events held by Presidents and their families in the executive mansion: the receptions, luncheons, dances, debut parties, weddings, christenings, children's parties and anniversary celebrations.

Each President entertains in his own select style—guided in a measure by custom, protocol and the mood of the era. Thus, through the years, White House hospitality has been by turns elegant, extravagant, sedate, stinted, formal, flamboyant, prim, puritanical, dignified and at times even rowdy and rambunctious. But, whatever the tempo, the glamor, excitement and prestige of being entertained in the President's house is always present.

An invitation to one of these events at the White House is the most coveted invitation in the nation. But invitations have not always been necessary to be among those entertained by the President. For more than a century public receptions—come one come all—were a tradition at the White House on certain days. Presidents and First Ladies stood for hours on these occasions shaking hands with the great, the near great and ordinary citizens who came to pay their respects, to see and be seen.

At a public reception on January 1, 1863, President Lincoln shook hands for three hours before going upstairs to his office to sign the Emancipation Proclamation. His arm was so stiff and numb, he feared his signature would be shaky, and someone would say he had compunctions about what he was signing.

After an exhausting stint of public handshaking, President Herbert Hoover abolished the traditional New Year's Day public receptions at the White House. Since then the coveted gold crested invitations issued in the name of the President or some member of his family have admitted only select thousands to the warmth and hospitality of the historic executive mansion.

Millions who cherish a hope of being a guest of the President of the United States in the White House some day may never have that privilege. This book is written for them, and also for those who would like to pattern their own entertaining after what is done in the White House.

It is a history of the entertaining done by Presidents and First Ladies at the White House down through the decades, beginning with John and Abigail Adams, who were the first host and hostess at 1600 Pennsylvania Avenue. It takes the reader to state dinners, receptions, luncheons and other parties. It tells little known events about these parties and the people who attended them. It describes table settings and decorations that may be adapted easily to home use. But more important to the hostess who wants to give extra flair and significance to her entertaining, it gives menus and recipes that were served by Presidents and First Ladies. This is a book to read, to keep, to use in planning your own social functions. It is also a book to give the hostess whose hospitality you enjoy.

M. S.

chapter 1

The Glittering State Dinner

Let's pretend the President of the United States has invited you to a state dinner at the White House.

You receive your invitation by mail about two weeks before the dinner. It is engraved with the Presidential seal in gold at the top and reads:

The President and Mrs. ....
request the pleasure of the company of
Your Name
at dinner
on (date)
at eight o'clock

The invitation comes in an expensive-looking white envelope with "The White House" stamped in gold in the upper left corner.

Inside, with the invitation, is a small sheet of white paper on which is typed in script the name of the person being honored at the dinner. This usually reads:

On the occasion of the visit of
His Excellency, the Prime Minister of ....
and Mrs. ....

Enclosed also is a small card engraved with this instruction:

Please send response to
The Social Secretary
The White House
at your earliest convenience

Enclosed, too, is another small card upon which is printed in pale, delicate ink, a South view of the White House—the same view that appears on a twenty dollar bill. Printed thereon is the following:

The White House
Admit at Southwest Gate
Your Name
Not Transferable
(*the date*)

There is no doubt that you will accept, and readily. An invitation to the White House is the closest thing in this country to a command appearance and only a very few people with unimpeachable excuses decline.

The main question is—what to wear? Some Presidents are more formal than others. At dinners given by President John F. Kennedy and his predecessor, Dwight D. Eisenhower, male guests wore white tie and tails; women wore floor-length formal dresses and above-elbow white kid gloves, their finest jewels and exotic hairdos. President and Mrs. Johnson took some of the starch out of state affairs when he set the example for other male guests by wearing black tie and dinner jacket, and she the standard for women by wearing the floor-length formal gowns and long gloves worn in preceding administrations.

Flowers are almost never worn by women to state occasions at the White House.

As a guest, you are one of approximately 100—more or less—invited to that particular evening's dinner, and although the invitation says eight o'clock, guests begin arriving at the Southwest gate of the White House fifteen to twenty minutes earlier. By sleek black limousine, taxicab and on foot they come up the wind-

ing drive between broad expanses of green to the South entrance of the executive mansion.

At the gate, your card of admission is checked by a guard. You are asked to show it again at the door to the mansion where a uniformed member of the White House police takes the card. If you wish to keep it as a souvenir, tell him, and it will be returned to you by mail. He will direct you through the oval diplomatic reception room on the ground floor of the mansion.

You go slowly through this beautiful and historic room for there is much here to see. If you are informed on the history of the White House, you know this is the room from which President Franklin D. Roosevelt delivered his famous fireside chats to the nation. Here, too, you pause to see the historic paper on the walls.

Called "Scenic America" and printed by Zuber & Co. in France in 1834, the wall paper shows five colorful views of America's natural wonders: Niagara Falls, the Natural Bridge of Virginia, Boston Harbor, West Point and New York Bay.

This paper was installed during Jacqueline Kennedy's tenure as First Lady. It was presented to the White House by the National Society of Interior Decorators. The paper hung originally in a private residence in Thurmont, Maryland. Just before the house was to be demolished to make way for a store, the paper was removed from the walls with a razor blade and sold for rehanging in the White House.

The room itself is furnished as a stately parlor of the late 18th century. The floor is covered with a huge oval rug of the Aubusson type woven with the seals of the fifty states. The dainty yellow-upholstered sofa and chairs are of New York and New England origin.

At the door to the adjoining China Room is a coat check and here you leave your wraps. It is a great improvement over the earlier years when guests left their cloaks in an upstairs hall unattended, and scrambled at the end of the evening to claim their own wraps before another guest made off with them. When you reclaim your wraps at the end of the evening, don't tip the checker. It isn't done in the White House.

If you wish, tonight you may pause in the China

Indians dance in the East Room in the 1960s following a dinner in honor of the President of Upper Volta. Students from Santa Fe, these Indians portray Gahan, mountain spirits revered by the Apache, wielding symbols of lightning.

Children's parties in the White House in the 1960s featured entertainment by costumed dancers.

Room to view the velvet-lined cases in which are displayed representative pieces of the china used by every President who has lived in the White House. The collection ranges from the delicate French Sevres china that Abigail Adams, the first hostess in the White House, brought with her from Philadelphia, to the handsome Castleton service plates with broad borders of gold ordered for the mansion by President and Mrs. Eisenhower. They were the last to buy china for the executive mansion.

In the softly carpeted corridor outside the China Room, uniformed, white-gloved military aides direct you up the broad marble stairway to the white-columned foyer. Here you see the source of the music you heard faintly in the lower hall. It is the Marine Band playing, each member resplendent in gold-buttoned scarlet coat and blue trousers.

In the foyer, you are greeted by a member of the White House social office, and another young military aide hands you a small white card showing a diagram of the table in the State Dining Room with an X marked where you will sit. Male guests also receive a card bearing the name of their dinner partner.

The military aide is one of 21 who serves day duty at a military post in the Washington area and serves voluntarily at social functions at the White House. He will escort you down the red-carpeted hall to the East Room at the left. At the door your name is announced to the guests already assembled. If you are the first arrival, your name is announced to the empty room.

You are not left, however, to your own resources. The aide takes you to the exact spot in the room where you are expected to stand and introduces you to those nearest you in the line which forms a half-circle around the room.

From nowhere it seems, a butler appears with a tray of drinks and you may take your choice—orange juice, tomato juice or one of several kinds of mixed drinks. You watch the arrival of other guests and talk to those around you. If you are observant, you will surmise that the guests are lined up alphabetically, unless they hold positions of rank in the Administration or titles in another government. Then they are lined up according to protocol.

Before this ceremony begins, a member of the White House staff has checked to see that every guest has arrived who accepted the invitation. If someone is missing, a hurried telephone call ascertains where he is and how soon he will arrive. After the guests have all arrived—except in rare cases where a delay has been unavoidable—the President and First Lady make their appearance.

Suddenly the Marine Band strikes up "Hail to the Chief." Above the music a rustle is heard on the stairway that rises from the foyer just outside the East Room door. On the steps are the President and First Lady. With them are a visiting chief of state and his wife. All pose for sixty seconds as photographers, brought in from below, click cameras. Then a military aide clears the area and, preceded by a color guard, the four—President and First Lady, visiting chief and his lady, arrive at the door to the East Room.

In the gold and white ball room, brilliantly lighted by three huge crystal chandeliers, every eye is focused on the doorway. The President stands erect and smiling in his black and white dinner wear. The First Lady looks dazzling in a gown that is sparkling with bright jewels. The visiting chief of state and his wife are dressed accordingly, unless, of course, they choose to wear the traditional formal dress of their homeland.

Slowly the line of guests begins to move as those at the beginning of it are introduced, by a military aide, to the President of the United States. He in turn presents each guest by name to the visitor and on to the two ladies.

After you have shaken hands with the host, hostess and the visiting guests of honor, you are directed by aides through the Green Room, back into the red-carpeted hall and down to the State Dining Room to take your place at the table.

The President leads the wife of the visiting chief of state to the Dining Room and the First Lady follows with the visiting chief. They take their places at the table and everyone is seated for the first course to be served.

What is on the menu? You will find it printed on a gilt-edged card beside your plate. Tonight it is a feast truly fit for a king:

Salmon Mousse
Tomato and Cucumber Salad
Tournedos Heloise Roast Potatoes
Green Beans with Almonds
Assorted Cheeses Crackers
St. Honore Cake

The table is a work of art. It is C shaped and overlaid with a fine white damask cloth. Seated in the center on the back of the C is the President. His back is to the fireplace whereon is engraved in marble the words from a letter John Adams, the first President to occupy the mansion, wrote to his wife, Abigail, before she joined him there. It reads:

"I pray heaven to bestow the best of blessings on this house and all that shall hereafter inhabit it. May none but honest and wise men ever rule under this roof."

Looking down from a portrait hanging above the mantel is a brooding Abraham Lincoln who spent four turbulent years in the mansion.

Seated opposite the President at the table, and facing the portrait and fireplace is the First Lady with the guest of honor at her side. All four sit on high back chairs upholstered in gold fabric.

Decorating the table directly in front of the President and First Lady is the handsome French bronze-dore plateau which President Monroe purchased in France in 1817. Reflected in the mirror of the gold-bordered plateau is a loosely arranged bouquet of white and yellow flowers in a vermeil[1] container, and at either end of the plateau are gold candelabra with tall tapers. At each side of the candelabra is another vermeil bowl of flowers.

Spaced at intervals on the table are other round, low arrangements of flowers with low white tapers on either side of the gold containers.

The table is set with china purchased during the administration of President Harry S. Truman. The plates have a broad green band, and outside that a smaller gold band. The Monroe gold flatware is used,

[1] Silver or bronze gilt.

as are vermeil nut dishes, ash trays and salt containers.

The white and gold of the table matches the antique white paneled walls, the gold framed mirror at one end of the room, the gold wall sconces, and the elaborate gold chandelier hanging in the center of the room.

The President and First Lady are always served first along with the guests of honor. Then the butlers begin serving other guests. The butlers work in teams of two, with one team assigned to each section of eight guests. Their uniform is white tie and tails and they perform faultlessly.

For each White House dinner between 12 and 20 extra butlers are hired. They are under strict orders not to repeat what they see and hear. Pay for the part-time butlers, who are investigated by the Secret Service and undergo rigid health checks before being allowed to serve at the President's table, is three dollars an hour. They report at 6 p.m. for an eight o'clock dinner and usually work for four hours.

Although the hourly rate is about the same as they earn at private parties, the butlers earn less on evenings they serve at the White House. This is because the hours are longer at a private party and the hostess adds a tip of from $3 to $5. The White House never tips.

However, the butlers will tell you, if you can find one who will talk, that they turn down private parties to work at the White House because it gives them prestige.

Many Washington hostesses who never make it to a White House dinner boast that their butler serves there, and there is some justification for her bragging. It means he is the tops in his trade if he is selected to serve a visiting king or president.

After dinner the part-time butlers clear the dining room and wash the crystal. The regular White House serving staff counts and wraps the gold and silver service and checks to see that none of the precious tableware is missing.

Should some guest slip a butter knife or nut bowl into his or her pocket as a souvenir, it would be discovered missing within minutes after leaving the room, and would be retrieved.

to his involvement in events relating to national and international affairs.

In many instances it has been the President's best method of keeping in touch with people from all walks of life who are invited to the White House. If you are invited to a state dinner in the White House, it is because you have made some contribution to that particular administration's programs. Or your presence is requested because of your achievement in some area, and the President is proud to present you as an American citizen to this guest of honor from another country.

Rarely do Presidents invite their personal friends to official dinners at the White House. There are ample opportunities to include them at private parties and events.

A hedge-bordered garden in the 18th century style and named in honor of First Lady Jacqueline Kennedy, is the setting for a summer afternoon party given by President and Mrs. Lyndon B. Johnson.

John Adams

chapter 2

THE FIRST WHITE HOUSE RECEPTION

The President's House, as the White House was called in its early years, was opened to the public for the first time on New Year's Day, 1801.

On that day, President John Adams and his wife, Abigail, gave a public reception. It was their first large-scale entertaining since taking up residence in the mansion, still largely unfinished and unfurnished.

They had sent the furniture from the presidential residence in Philadelphia to the new Federal City but it was hardly adequate to fill a house as large as the new Executive Mansion.

John Adams was nearing the end of his four years in office when he arrived in Washington early in November, 1800, to prepare for the convening of Congress in the new Federal City later that month. He was followed by Abigail, then 56 and not in the best of health, on Sunday, November 16.

Citizens of the new capital had waited impatiently for them to open the new mansion so all might come and see it and be entertained. This New Year's Day reception was in response to their requests.

The guests were received in the oval library on the second floor, a handsome room commanding a fine view of the Potomac River nearby, and the rising Virginia hills beyond.

Glowing crimson curtains hung at the windows, and in the room were two mahogany sofas and twenty-two chairs, all upholstered in crimson. There was also a piano, a harp and a guitar—for Mrs. Adams was a brilliant and talented woman who loved music. The floor was brightened by an Aubusson rug and a huge fire crackled on the hearth, giving warmth to the guests who arrived shivering from the midwinter cold.

Here and there, lighted candles cast a cheerful glow on the scene, and the Marine Band, making its first appearance at the President's House, provided music.

Cheerful, genteel Abigail, who had a warmth and grace that was lacking in her sometimes austere, aloof husband, remained seated to receive the guests. She was gowned elegantly in brocade and velvet. At her side stood the President, wearing a powdered wig, knee breeches, gay colored waistcoat, high stock collar, silk stockings and silver buckles on his slippers.

No hands were shaken, following the custom of George Washington when he, as President, received guests in New York and Philadelphia. As the guests stepped forward, their names were announced. Presi-

The elite enjoy a dancing party at the Presidential mansion during the John Quincy Adams Administration.

dent Adams bowed to each and repeated the names. Mrs. Adams smiled and nodded. Her natural faculty for cordiality tended to put everyone at ease.

After paying their respects, the guests passed on to be served refreshments by a waiter. There were many goodies—cakes and tarts, curds and cream, tea, coffee, punch, syllabub, floating island, trifle and many other sweets of that period.

One of the refreshments served at her reception in the President's House on January 1, 1801, was Floating Island. Here is a recipe for it:

FLOATING ISLAND

1 quart sweet milk
5 egg yolks
5 egg whites
8 tablespoons sugar
1/8 teaspoon salt

Beat together five egg yolks and one white. Scald milk and stir a little into the egg mixture to prevent curdling, then stir in all the milk and five tablespoons of sugar. Cook over slow fire until thickened, remove, cool and flavor with vanilla. Pour into glass bowl and let custard chill. Add salt to remaining whites of four eggs and whip to a froth, adding three tablespoons sugar. Pour froth on a shallow dish of boiling water and let stand until cooked by the steam passing through. When firm, drop a tablespoon at a time on top of the custard and far enough apart so the white islands do not touch. Serve cold.

Mrs. Adams had spent days preparing for this event, which started a White House tradition that lasted until abolished by President Hoover 128 years later. Much of the time during the preparation for the reception, her heart was heavy. She was disappointed in the President's house; her husband gloomy and unhappy with the turn of political events. And their second son, Charles, had died in poverty only a few weeks after she had visited him in New York.

Her disappointment in the President's house was revealed in a letter written to her sister five days after her arrival on that memorable Sunday. It had been a tiring journey for her and her entourage, that included ten horses and nine persons. One of them was

Suzannah, the four-year-old daughter of her dying son, Charles.

Enroute from Baltimore, they got lost in the woods and went nine miles on the road to Frederick, Maryland, before retracing their path and getting onto the road to Washington.

Telling her sister about what she found, Mrs. Adams wrote:

"I arrived about one o'clock at this place known by the name of "the city" and the name is all that you can call so! As I expected to find it a new community with houses scattered over a space of ten miles, and trees and stumps in plenty, with a castle of a house, so I found it.

"The President's House is in a beautiful situation in front of which is the Potomac with a view of Alexandria. The country round is romantic but a wild wilderness at present.

"But surrounded with forests can you believe that wood is not to be had, because people cannot be found to cut and cart it! Breisler entered into a contract with a man to supply him with wood . . . but a few cords only has he been able to get. We have had some very cold weather and we feel it keenly. This house is twice as large as our meeting house. I believe the great hall is as big. I am sure 'tis twice as long. Cut your coat according to your cloth. But this house is built for ages to come. The establishment necessary is a task which cannot be borne by the present salary. Nobody can form an idea of it but those who come into it. . . . Not one room or chamber is finished of the whole. It is habitable by fires in every part, thirteen of which we are obliged to keep daily, or sleep in wet and damp places. To assist us in this great castle, and render less attendance necessary, bells are wholly wanting, not one single one being hung through the whole house, and promises are all you can obtain. This is so great an inconvenience that I know not what to do!. . . .

"The ladies from Georgetown and in the "city" have many of them visited me. Yesterday I returned fifteen visits[2]—but such a place as Georgetown!. . . . It

[2] Abigail called on fifteen households, whose members had first paid their respects to her in the President's House.

is the very dirtiest hole I ever saw for a place of any trade or respectability of inhabitants. It is only one mile from me, a quagmire after every rain. Here we are obliged to send daily for marketing. The capitol is near two miles from us. As to roads, we shall make them by the frequent passing before winter! But I am determined to be satisfied and content, to say nothing of inconvenience, etc. That must be a worse place than even Georgetown that I could not reside for three months! If they will put me up some bells and let me have wood enough to keep fires, I design to be pleased. We have not the least fence-yard or other convenience without, and the great unfinished audience-room I make a drying-room of, to hang the clothes in. The principal stairs are not up, and will not be this winter.

"Six chambers are made comfortable; two are occupied by the President and Mr. Shaw (his secretary); two lower rooms for a common parlor, and one for a levee room. Upstairs there is the oval room which is designed for the drawing room and has the crimson furniture in it. It is a very handsome room now; but when completed, it will be beautiful. If the twelve years in which this place has been considered as the future seat of government had been improved, as they would have been in New England, very many of the present inconveniences would have been removed."

If the Adamses' hospitality at that first New Year's Day reception lacked some of the brilliance and warmth of their social functions in Philadelphia, they could hardly be blamed.

Mrs. Adams complained that many of their possessions were stolen and many more broken during the move, including more than half of her tea china. She found it very difficult to get replacements in Georgetown.

Yet she was a gracious hostess who had seen "the luxury and pomp of state, the power of riches and the influence of titles" during the years she was in Europe with her husband when he was commissioner to France and minister to Holland and England.

Adams had served as vice president for eight years under George Washington. During that time he and Abigail had lived in New York, and later, Philadelphia, when the presidency was moved there in 1790.

It was Adams who had helped Washington draw up a system of etiquette for the presidency. It specified that the President would issue dinner invitations only to official persons and strangers of distinction. President Washington held public levees,[3] mainly stag affairs, every Tuesday between three and four o'clock.

Surrounded by Adams and members of his cabinet, he stood in the large dining room of the President's Palace in New York and bowed as the guests were introduced by his secretary. He exchanged a few words with each, and afterwards, as they were leaving, they bowed to him in salute again.

Washington held a "congressional" dinner every Thursday at four o'clock and private dinners other days at three o'clock.

Mrs. Washington was at home to callers every Friday night at eight. President Washington usually

[3] A reception or assembly of callers or guests; especially a reception given by the President of the United States.

attended these events, which drew men as well as women. As Mrs. Washington received, Abigail Adams had a place of honor at her right.

Lafayette, Thomas Jefferson, and Benjamin Franklin attended, along with other government officials, and occasionally their wives.

When it came John and Abigail Adams's time to occupy the presidential residence in Philadelphia, they gave a formal dinner once a week for from 15 to 20 guests, although Abigail found these dinners very expensive considering her husband's salary. But she recognized their worth and wrote "More is to be performed by way of negotiations many times at one of these entertainments, than at twenty serious conversations."

George Washington

The Adamses served this at the President's residence in Philadelphia:

ADAMS' FOURTH OF JULY PEAS

½ cup water
2 cups small green peas
2 cups fine shredded lettuce (loosely packed)
1 teaspoon grated onion
1 tablespoon butter
salt

Melt butter in sauce pan. Add grated onion and shredded lettuce. Saute lightly. Add peas. Add water and cook rapidly for about 25 minutes. Drain off any excess water, add another tablespoon of butter, and shake pan over heat until the butter is distributed and excess moisture is gone.

Like the Washingtons before them in Philadelphia, President Adams held a weekly levee and Mrs. Adams a weekly reception. At her last reception in that early capital, some 200 guests came, and the rooms were so crowded that the air quickly became oppressive. The Adamses were nearly exhausted when it was over.

In Washington, they did not feel the need to entertain so frequently. In this city of few homes, many Congressmen lived in boarding houses and left their wives at home.

One of the first parties the Adamses gave in the Executive Mansion was a pre-Christmas affair for members of Congress and the few wives who were in the City. This party was something less than a social success, but no fault of the Adamses.

Before the party, Abigail burned 20 cords of wood in an effort to dry and warm the rooms. But they were still damp and cold when the guests arrived, and they sat around trying to conceal their discomfort. This did not improve the situation, however, and everyone left early.

Added to the chill of the rooms was the attitude of some members of Congress toward her husband. They made fun of him for his coolness and preference for the pomp and ceremony he had known in Europe, and nicknamed him "Nonny Johnny Adams."

Another Christmas season party the Adamses gave in the White House was more of a success as far as the physical comfort of the guests, but it, too, ended on a discordant note. It was for their four-year-old granddaughter, Suzannah. The young guests kept warm by playing games. But during one of the games a little girl broke one of Suzannah's prize possessions, a piece of her doll's tea set. She retaliated by breaking the nose of the guest's doll and it took the President himself to settle the quarrel that ensued.

After the New Year's Day reception, the President and Mrs. Adams did little formal entertaining. Abigail had some ladies in to dine one day, setting the table with her fine white Sevres china decorated with blue cornflowers. And on February 7th, she and the President gave a dinner for the judges and heads of departments and their ladies. Six days later she departed from the capital to return to the Adams' house in Braintree (now Quincy), Massachusetts. Adams joined her there in March.

Though she had performed her role of First Lady well and set a pattern for all who would follow her in it, Abigail Adams, the daughter of a New England minister who was educated in her own home, was happy to return to the quiet farm in Massachusetts which she ran as a profitable business. It was here she had reared her children.

Thomas Jefferson

chapter 3

A New Pattern of Entertaining

Thomas Jefferson brought a fresh and original pattern of Presidential entertaining. Soon after he took office on March 4, 1801, he decided to dispense with the weekly levees and drawing rooms held by his two predecessors. Instead, he decided to hold two public receptions a year: one on New Year's Day and the other on July 4, the nation's birthday.

This decision greatly displeased the ladies of Georgetown, who looked forward to going to the President's House with the 57-year-old widower from Virginia as the host. They plotted ways to induce him to resume the former pattern of presidential entertaining.

At two o'clock one day they called at the mansion in a group, as if a drawing room had been announced. They were shown to the reception room where they waited an hour before Jefferson returned from his daily ride on horseback. An aide told the President that the reception room was full of guests. The President had heard the gossip and knew what was up. He entered the room still wearing his dusty riding clothes and greeted the ladies warmly. He pretended it was a pleasant accident that they should all call at the same time. His charm quickly won them and they lacked the heart to spoil it by forcefully demanding a resumption of the weekly levees.

Jefferson listened to their mild request and then explained his position, which the ladies accepted with good grace.

Jefferson had seen Washington and Adams stand stiffly at one end of a reception room each week and greet the scores who were introduced to them. He was determined that he would not spend his time in such a manner.

However, he probably entertained more often in the White House than has any other President. He had an informal dinner for twelve to fourteen guests, including himself and his secretary, almost every day of the week.

The guests, invited for "half after three," sat down to eat at four o'clock and rose at six. They went from

The Red Room which has been the setting of many White House parties, as furnished today.

the dining room to an oval sitting room for coffee, and then dispersed.

Jefferson's hospitality was lavish, with a free and easy character. He recognized the value of the dinner table for political negotiations, and invited the persons he wanted to see at the moment, rather than following any list in sequence or form of protocol.

These cherished recipes are indicative of the fine food he served:

JEFFERSON'S FILLED PANNEQUAIQUES

12 5 to 6 inch crepes (make ahead and cool)
2 cups diced cooked chicken
1 cup sliced fresh mushrooms
¼ cup minced onion
¼ pound butter
2 tablespoons dry sherry
1 egg yolk
½ cup milk
pinch of salt

Melt butter in pan and saute minced onion until barely glazed. Add mushrooms and saute slowly for five minutes (low heat, do not let onion get brown). Add chopped chicken. Cool and add milk and sherry. Last, stir in the yolk of one egg.

Fill each crepe with about 2 heaping tablespoons of the mixture and roll. Trim ends of crepes so that they are even. Arrange in a row on a platter.

Mix together:
½ cup hollandaise sauce
½ cup cream sauce

Fold in:
1 cup whipped cream

Spread this sauce thickly over the crepes, allowing some sauce to cover ends. Dust with grated parmesan cheese, and place under broiler until sauce is delicately browned and bubbling. Serve immediately. Serves six.

ASPARAGUS WITH MONTICELLO SAUCE VINAIGRETTE

Put 1/4 cup olive oil or sesame oil, 1/2 cup vinegar, 1 teaspoon salt, 1/4 teaspoon freshly ground black pepper, and a dash of cayenne in a bowl and beat well with hand beater. Add 1 tablespoon each of chopped parsley, pimiento, and green pepper. Add 1 teaspoon each of chopped tarragon and chervil. Serve sauce over cooked and chilled asparagus on Bibb lettuce. Add 1 tablespoon capers, if desired.

BLACK WALNUT MERINGUES

Beat two egg whites and 1/8 teaspoon salt with wire whisk until fluffy but not dry. Egg whites should hold up in peaks when whisk is lifted. Add 1 cup sugar slowly while beating constantly. Fold in 1 teaspoon vanilla and 1 cup chopped black walnuts. Drop by teaspoonful on waxed paper or cookie sheet. Bake in slow oven, 225 degrees, until set and partly dry.

There seemed to be no standard form for the wording of his invitations. Sometimes his dinner invitations read: "Th. Jefferson requests the favor of Mr. and Mrs. Smith to dine with him Tuesday, at half after three and any friends who may be with them. April 25, 1803. The favor of a reply is requested."

Another time the invitation might read: "Th. Jefferson presents his compliments to Mr. Smith and requests the favor of his company to dinner." Added at the bottom, in the left, or center, was "The favor of an answer is desired."

A widely-traveled man-of-the-world who had lived in Paris from 1785-1789 while he was treaty commissioner to France, Jefferson was an epicure with a penchant for French and Italian foods. He delighted in importing foreign delicacies for his table, and a French steward, Etienne Lemaire, headed his staff at the White House.

He spent money freely to set the most lavish table in Washington and it was not unusual for Lemaire to spend fifty dollars upon a single day's marketing in Georgetown. During his first year in the White House,

Jefferson spent $4,504.84 for provisions; $2,003.71 for groceries; $2,797.28 for wines; and $2,675.84 for servants, his accounts show. His bills for wine alone totaled nearly $11,000 during his eight years in the Presidency. The President's salary in those days was $25,000 a year.

One guest who dined with Jefferson on Saturday, February 6, 1802, recorded they had "rice soup, round of beef, turkey, mutton, ham, loin of veal, cutlets of mutton, fried eggs, fried beef, a pie called macaroni . . . ice cream . . . and many other jim cracks, a great variety of fruit, plenty of wine and good."

Jefferson is credited with introducing ice cream at the White House, as well as many other imported delicacies.

CITRON CREAM

Cut the finest citron melons when perfectly ripe. Take out the seeds and slice the nicest part into a china bowl in small pieces that will lay conveniently. Cover them with powdered sugar and let them stand for several hours. Then drain off syrup they have made and add to as much cream as will retain the strong flavor. Then freeze it. Pineapples may also be used in this way.

CABBAGE PUDDING

Shred 1/2 pound of lean beef and a pound of suet very fine, add the yolks of three eggs, one spoonful grated bread, some sweet herbs, pepper, salt and onion. It will fill a cabbage which must be parboiled and opened at the top. Scoop it out till you think it will receive the meat. Fill it, close it up, tie it hard and close in a cloth. Place in a pan of water. When it has boiled a little, tie closer. It must boil two and one-half hours.

GOOSEBERRY FOOL

Pick the stems and blossoms from two quarts of green gooseberries. Put them in a stewpan with their weight in loaf sugar and a very little water. When sufficiently stewed pass the pulp through a sieve, and when cold add rich boiled custard till it is like thick cream. Put it in a glass bowl and lay the brothed cream on top.

Another dinner given by Jefferson was described by Mrs. Harrison Smith, wife of the publisher of *The National Intelligencer,* as follows:

"Mr. Smith and I dined at the President's. He has company every day but his table is seldom laid for more than 12. This prevents all form and makes the conversation general and unreserved. One circumstance though minute in itself has a great influence on the conversational powers of Mr. Jefferson's guests. Instead of being arranged in straight parallel lines where you cannot see the countenance of those who sit at the same table, the circle or round oval table where all could see the others faces and feel the animating influence of looks as well as words. Let any dinner-giver try this experiment and he will certainly be convinced of the truth of the fact. The small, well assorted company seated around a circular table will insure more social enjoyment than any of the appliances of wealth and splendor without any concountenance."

The White House in Jefferson's day, as now, had two dining rooms on the main floor. The large one then was on the northwest corner and the small one in the southwest corner (it is just the opposite today). It was in the smaller dining room that Jefferson was host.

The room was furnished elaborately with an elegant sideboard, a large mahogany dining table in six pieces, a small dining table in three parts, two glass cases to contain the silver and plated ware, an oval breakfast table, and fifteen chairs in black and gold. The windows were hung with chintz curtains, and on the walls, two elegant girandoles[4] and two looking glasses. Before the table was set each day, Jefferson had a canvas cloth painted green laid over the floor to protect it from grease and crumbs. The cloth was removed at the end of the meal.

Jefferson invented his own device to keep servants from overhearing the conversation at the table. This device was a "dumb waiter,"—a stand with shelves containing the food to be served. This was placed between the guests, who could then serve themselves. Later, he designed circular shelves in the wall that

[4] A branching chandelier or bracket light.

could be filled with dishes on one side by servants, and turned around, where the guests removed them, serving themselves, on the other side.

When he moved into the White House, Jefferson had been a widower for 18 years and he kept his promise to his young wife that he would never remarry. His bachelor's life may have accounted in part for his independence in dress. He refused to wear a wig after 1800, although most men did; and he wore pantaloons, not breeches and stockings. One Ambassador was shocked when he presented his credentials to President Jefferson to find him wearing pantaloons and slippers.

The White House was without a full-time hostess during Jefferson's eight years as President, but he made out very well without one. His two daughters, Martha, who was married to Thomas Mann Randolph, and Mary, who was married to John Wayles Eppes, came occasionally to Washington to serve as their father's hostess. But more often he called on the warm and gracious Dolley Madison, wife of his Secretary of State, to fill the hostess role.

One guest of Jefferson's wrote how Dolley Madison conducted her through the rooms of the mansion, taking her from room to room, including the chamber occupied by the President and his secretary. She even opened up the President's wardrobe and showed his device, similar to a turnstile, for hanging up jackets and breeches.

On another occasion when Dolley was serving as hostess and the guests included the British Minister and Mrs. Merry, Jefferson's "pell-mell" system of etiquette created an international incident. When dinner was announced, the President offered his arm to Mrs. Madison, and Secretary of State Madison in turn offered his arm to the lady nearest him, and other cabinet members followed suit. Minister Merry, a gentleman of the old school who had served many years at the court in Madrid, was bewildered when he found himself left to go in with his wife! Dolley had whispered to Jefferson, "take Mrs. Merry" but he didn't heed her. Immediately after dinner, Merry sent for his carriage and left. He made a formal complaint to his government and refused to dine with either Jefferson or Madison again.

A diplomat at the dinner predicted the incident would "be the cause of war," and it is true that while Madison occupied the White House as President, the British marched on Washington and burned the mansion. But the Merry incident is not credited with contributing to this later development.

President Jefferson held his first public reception at the White House on July 4, 1801, and the like of it the new Federal City had never seen.

The public was invited to come at 12 o'clock noon, and, as the great East Room of the mansion was still unfinished and unfurnished, Jefferson received in the drawing room (called the Blue Room today) which he had recently carpeted and furnished. He abolished the old practice of bowing and held his hand outstretched in democratic fashion to clasp the hands of his guests. They were surprised but pleased at this form of greeting.

Since Congress was not in session, the guests numbered only about 100 and included all the public officers, members of the diplomatic corps, most of the respectable citizens of the city and a few distinguished strangers. Many were shocked to find the President surrounded, as he received, by five strong, silent Cherokee Indian chiefs, who were in Washington to celebrate a new treaty of friendship.

Jefferson was pleased with the sensation the chiefs created, each wrapped in a blanket and wearing deerskin moccasins and cloth-ornamented leggings with embroidery and fringes of colored beads. Their faces were covered with paint and their heads shaven except for a tuft of hair. Jefferson had requested that they appear in their native dress, but he left up to the wife of the Secretary of War the attire for the squaws.

After consultation with other cabinet wives, she dressed the squaws in brightly colored chintz without any trinkets or ornaments. One cabinet wife placed a bunch of artificial flowers in the hair of one of the Indian ladies, who removed it immediately and carried it to her husband and stuck it behind his ear. He let it remain there.

The Marine Band, which had marched up Pennsylvania Avenue just before the reception began, saluted the house, marched inside, and took its place in the hall. It then struck up a new tune, "The President's March." As the band continued to play, it paraded through the state rooms of the mansion. Many guests trooped behind them, until the musicians once again resumed their places in the hall.

President Jefferson talked easily and cheerfully with his guests and then invited them into the State Dining Room, where four large tables were covered with refreshments.

Here are recipes for some of the refreshments served by Jefferson to his guests on various occasions:

SPONGE CAKE

1 dozen eggs, whites and yolks beaten separately
1 pint sugar
1 pint flour
juice and grated rind of one lemon

Mix yolks, sugar, flour and lemon. Add whites. Pour into cake mold and bake in moderate oven.

SPICED CIDER

Mix one quart of sweet cider with one-fourth cup of sugar, a pinch of salt, eight short pieces of stick cinnamon, twelve whole cloves and eight whole allspice. Bring to the boiling point, cool and let stand for several hours. Reheat, remove the whole spices and serve hot with cookies.

OLD VIRGINIA COCKTAIL

Mix one cup of grape juice, one cup of water, one half cup of orange juice, one fourth cup of lemon juice with one cup of cider. Stir and pour over crushed ice. Serves six.

Among the diplomats was Tunisian Minister Meley Meley and his entourage dressed in silk robes, tall turbans and slippers with upturned toes, which they removed at the door in order not to defile the floors of the President's house.

Minister Meley was fascinated by the Indian chiefs and took off his turban to show them his head was shaved just like theirs.

After the guests left at 3 o'clock, Jefferson entertained the Indians at dinner. They had remained calm and dignified during the reception but at dinner their stony facade broke when wine was placed on the table in coolers filled with ice. The Indians looked at each other with expressions of doubt and surprise. One picked up a piece of ice, felt it and passed it to the others. They began talking rapidly to each other. When Jefferson asked for an interpretation, he was told one chief said, "We now believe that what our brothers told us when they came back from the great cities was all true, although at the time we thought they were telling us lies when they told of all the strange things they saw, for they never saw anything so wonderful as this that we now see and feel. Ice in the middle of summer. We can now believe anything."

Year after year the July 4 receptions increased in size until it was necessary to move them onto the South Lawn where colorful tents and booths were erected. Here troops marched in formation and the President, bareheaded, stood on the steps with the heads of the departments and the foreign ministers, to return the salutes. After the parade, the soldiers were dismissed to join the crowds on the lawn while the officers went inside the mansion with the President. There the rooms were filled with citizens of wealth and position.

The President stood in the middle of the oval drawing room and shook hands with all who wished to come by. Tables in the dining room were filled with confectionaries, wines, punch, lemonade and other refreshments. And the Marine Band played from its traditional station in the hall.

The July 4 reception in 1803 was especially colorful because it was a double celebration. There were banners, bands, speeches and cheers, for Jefferson announced the acquisition of the Louisiana territory.

Jefferson's first New Year's Day reception was a gala event because the weather was fine, and, as Congress was in session, the number of guests was greater than at his midsummer event. He was joined on this occasion by his two daughters, Mrs. Madison and her sister.

Mrs. Madison was described by one eyewitness as wearing, "A turban of white satin with three large white ostrich feathers hanging over her face, very becoming indeed. Her dress, too, of white satin, made high in the neck with long sleeves and large capes trimmed with swan's down, was rich and beautiful. She never wears diamonds in the morning; she looks remarkably well, and as much like a bride as a queen. . . . Her stately person, her lofty carriage, her affable and gracious manner, would make her appear to advantage at any court in the world. . . . She moves like a goddess and she looks like a queen."

President Jefferson loved having his grandchildren in the White House and one year had all six of them with him there at Christmas. He was so delighted he gave a big children's party for them with Dolley Madison acting as hostess.

Some 100 children were invited and they had a wonderful time playing games and eating the food that Dolley had planned for them. The President joined in the fun showing the children his pet bird that perched on his shoulder, took food from his lips and walked up the stairs with him. He also got out his violin and played for the little ones to dance.

After his administration, Jefferson went on to perhaps the most ambitious effort of his life: the establishment of a university that would live up to his ideal of equal opportunity for all. He is noted in his dealings with men for his earnestness of manner and certainly as one of the most genuinely unpretentious hosts ever to inhabit the White House.

Keeping a tradition started in 1801, musicians from the Marine Band play at a White House party for children at Christmas time, in the 1960s. Mrs. Dean Rusk, wife of the Secretary of State, listens with the children.

James Madison

chapter 4

The Madison Era

Dolley Madison's reign as first lady was regarded by many as the golden age of social life in the White House. The gay, glamorous, gracious Dolley set the pace with her own parties, and was looked upon as a queen when she went to parties given by others.

President Madison, who was studious, retiring and small in stature (only five feet four inches tall), faded into the background when he and Dolley entertained. But Dolley, who stood five feet nine inches and almost always wore high heels and a towering turban, was visible wherever she wandered.

At Dolley's suggestion, Madison restored some of the ceremony and protocol of the Washington and Adams administrations. Much to the delight of the citizens of the capital city, he and Dolley held a drawing room each Wednesday evening.

Crowds came not only from Washington but from all over the nation to join in the blazing splendor of Dolley's drawing rooms. And Dolley, who loved people and parties, entertained frequently at dinners and receptions also.

It rained the night of her first reception in the White House, but she had the mansion glowing with lamps. Some 200 people turned out to be received by the new President and his wife in the oval drawing room that she had redecorated.

She had thrown out the faded red, and hung the windows with bright yellow damask. She had bought new sofas and chairs in the same sunny yellow, and hung a large emblem of the noonday sun above the mantel before which she stood to receive her guests.

She often wore yellow to match the decor, or red to contrast with it, and give added brightness to the occasion.

After greeting everyone, Dolley went from room to room talking to first one delighted guest, then another. Often she was seen with a book in her hand. She was not really a serious reader, but the book served as a topic for conversation if she could think of nothing else.

Her Wednesday night drawing rooms became the rendezvous for political leaders of both parties, and they could rest assured that gracious Dolley would avoid uncomfortable topics of conversation.

Sometimes the crowd was so large on Wednesday nights that the Madisons would hold a second reception on the following night.

Washington Irving attended one of her receptions and described it as follows:

"I emerged from dirt and darkness into the blazing splendor of Mrs. Madison's drawing-room. Here I was most graciously received; found crowded collection of great and little men, of ugly old women and beautiful young ones, and in ten minutes was hand in glove with half the people in the assemblage. Mrs. Madison is a fine, portly, buxom dame who has a smile and a pleasant word for everybody. Her sisters, Mrs. Cutts and Mrs. Washington, are like the two merry wives of Windsor; but as to Jemmy Madison—ah poor Jemmy! He is but a withered little apple-John."

William Campbell Preston, son of one of Mrs. Madison's friends in Philadelphia, paid a visit to the Madisons at the White House and wrote in his journal as follows:

"I and my conductor proceeded in the hack in utter silence. The appearance of the house and grounds was very grand. There was a multitude of carriages at the door, many persons were going in and coming out; especially many in gaudy regimentals. Upon entering the room where there were 15 or 20 persons, Mr. Madison turned toward us, and the General said in presenting me, "My young kinsman, Mr. Preston, who has come to present his respects to you and Mrs. Madison."

"The President was a little man with powdered head having an abstracted air and a pale countenance, with but little flow of courtesy. Around the room was a blaze of military men and naval officers in brilliant uniforms. The furniture of the room with the brilliant mirrors, was very magnificent. While we stood, Mrs. Madison entered—a tall, portly, elegant lady with a turban on her head and a book in her hand. She advanced straight to me, and extending her left hand, said: 'Are you William Campbell Preston, the son of my old friend and beloved kinswoman, Sally Campbell?'

"I assented. She said 'Sit down, my son, for you are my son, and I am the first person who ever saw you in this world. Mr. Madison, this is the son of Mrs. Preston who was born in Philadelphia.' The President shook hands with me cordially."

Another guest at one of Dolley's receptions wrote of her: "I would describe the dignified appearance of Mrs. Madison but I could not do her justice. 'Tis not her form, 'tis not her face, it is the woman altogether, whom I should wish you to see. She wears a crimson cap that almost hides her forehead but which becomes her extremely and reminds one of a crown from its brilliant appearance, contrasted with the white satin folds and her jet black curls; but her demeanor is so far removed from the hauteur generally attendant on royalty that your fancy can carry the resemblance no further than the headdress."

Dolley often appeared at her parties wearing low-cut dresses without the handkerchief generally worn around the neck to cover bare shoulders. On one such occasion when a Quaker friend from Philadelphia was a guest, she raised her wine glass to her lips and said gaily: "Here's to thy absent broadbrim, Friend Hallowell." He quickly replied, "And here's to thy absent kerchief, Friend Dolley."

Although Madison was very dignified and reserved in the reception rooms, toward the end of some evenings with friends he became anecdotal, facetious and occasionally risque. His wife never failed to rescue the situation with tact, poise, and a sense of humor.

During the summer of 1812, the Madisons often entertained Indians at the White House. Forty of them from different tribes gathered there one evening, and the dinner speeches continued for nearly six hours. When Mrs. Madison retired to her chamber, she looked into the mirror, and saw one of the visiting red men, painted and feathered, in her room. She calmly rang for a servant who persuaded the Indian to leave.

The New Year's Day receptions held by the Madisons were as brilliant as any ever held in the executive mansion. Mrs. William Seaton, wife of the editor of The National Intelligencer, wrote this account of the one on January 1, 1814:

" . . . yesterday being New Year's day, everybody, affected or disaffected towards the government, attended to pay Mrs. Madison the compliments of the season. Between one and two o'clock we drove to the President's where it was with much difficulty we made good our entrance, though all of our acquaintances endeavored with the utmost civility to compress themselves as small as they could for our accommodation. The Marine Band stationed in the ante-room, continued playing in spite of the crowds pressing on their very heads. But if our pity was excited for these hapless musicians, what must we not have experienced for some members of our sex, who, not foreseeing the excessive heat of the apartments, had more reason to apprehend the efforts of nature to relieve herself from the effects of the confined atmosphere. You perhaps will not understand that I allude to the rouge which some of our fashionables had unfortunately laid on with an unsparing hand, and which assimulating with the pearlpowder, dust and perspiration, made them altogether unlovely to soul and to eye.

"Her majesty's appearance was truly regal—dressed in a robe of pink satin, trimmed elaborately with ermine, a white velvet and satin turban, with nodding ostrich plumes and a crescent in front, gold chain and clasps around the waist and wrists. 'Tis here the woman who adorns the dress, and not the dress that beautifies the woman. I cannot conceive a female better calculated to dignify the station which she occupies in society than Mrs. Madison—amiable in private life and affable in public, she is admired and esteemed by the rich and beloved by the poor.

Dolley Madison

"You are aware that she snuffs, but in her hands the snuff-box seems only a gracious implement with which to charm. Her frank cordiality to all guests is in contrast to the manner of the President who is very formal, reserved and precise, yet not wanting in a certain dignity. Being so low in stature, he was in danger of being confounded with the plebian crowd and was pushed and jostled about like a common citizen—but not her ladyship. The towering feathers and excessive throng distinctly pointed her station wherever she moved.

"After partaking of some ice creams and a glass of Madeira, shaking hands with the President and tendering our good wishes, we were preparing to leave the rooms when our attention was attracted through the window towards what we conceived to be a rolling ball of burnished gold, carried with swiftness through the air by two gilt wings. It stopped before the door and from it alighted, weighted with gold lace, the French Minister and suite. We now also perceived that what we had supposed to be wings were nothing more than gorgeous footmen with chapeaux bras, gilt braided skirts and splendid swords. Nothing ever was witnessed in Washington so brilliant and dazzling—a meridian sun blazing full on this carriage filled with diamonds and glittering orders and gilt to the edge of the wheels—you may well imagine how the natives stared and rubbed their eyes to be convinced it was no fairy dream."

The lavish hospitality of the Madisons made its greatest showing at their late afternoon dinners in the White House. Guests came at four o'clock and found themselves seated at tables groaning with fine French dishes and delicious desserts.

Mrs. Madison had acquired from the British embassy a Parisian steward, Jean Pierre Sioussat, who became her master of ceremonies and was always at her side. She had 30 other servants who carried out the orders and duties prescribed by French John, as they called him. She hired extra servants from neighboring plantations for thirty-five cents an evening when needed, and had one waiter for each guest at her dinners.

Mrs. Madison presided at the head of the table with guests on right and left. The President sat at the side, thus sparing himself the effort of serving, or leading the conversation. But with Dolley's quick conversational abilities, the dinners were always a success.

Their success was assured also by an abundance of good food, superb service, and the purposely light tone of conversation.

Following the main courses, which often included waterfowl, deer, game birds and oysters as well as ham and beef, (all served with vintage wines), the table was set with ice creams, macaroons, preserves and various cakes. These were followed by almonds, raisins, pecan nuts, apples, pears and other fruits.

Candles were lit just before the ladies left the table. And while they repaired to the drawing room for coffee, the gentlemen continued at the table half an hour longer to enjoy a social glass.

Left among her papers were some of her treasured recipes. Here are some of them:

GINGER POUND CAKE

2 lbs. flour	1 pint molasses
1 lb. sugar	½ pint sour cream
1 lb. butter	1 cup ginger
10 eggs	1 teaspoon baking soda (dissolved in warm water)

Mix and bake as a pound cake.

WHITE NUT CAKE

1 cup sugar	2 cups flour sifted three times
½ cup butter	2 teaspoons baking powder
1 cup milk	1 cup chopped floured nuts
3 egg whites	½ teaspoon vanilla

Bake in a loaf pan.

STRAWBERRY ROLL

To one quart of berries add one cup of sugar and one-half cup flour. Line a bread pan with short biscuit dough. Fill with the berries and cover over, folding the bottom crust over the berries. Pinch the top so that the steam can escape. Bake for 20 minutes.

Dolley Madison extended her hospitality to the President's official conferences. She often sent her famous seed cakes and bouillon in to refresh the conferees. And while the President held his cabinet meetings, she invited the members' wives to "dove parties" at which they kept up with events of the day and discussed projects that were uppermost in Dolley's life at the moment.

She was not a frivolous person, despite her love for parties. She sponsored many worthy causes as First Lady and contributed to many charities. She helped raise money for the Washington Monument, and led the women of Washington in equipping the Lewis and Clark expeditions when they were sent out by President Jefferson.

Despite her plain childhood and strict upbringing in a devout Quaker family, Dolley was well prepared for her role as First Lady. She may have dreamed of such a life when she was a little girl in Virginia. We know that she was given some family trinkets by her grandmother, which were such a joy to her that she hid them rather than risk forfeiting them as frivolous vanities. She sewed them into a little bag which she hung around her neck under her somber dresses.

The story is told that the greatest calamity of her childhood occurred when the string of the bag broke and her treasures were lost in the woods as she walked to school.

Dolley was a widow with one son when she met and married James Madison, a wealthy Virginian who was serving in Congress in Philadelphia. As a spokesman for his party, his home was open for entertaining during the entire Congressional season.

During John Adams' term in the presidency, however, she stayed at home in Virginia and arrived in Washington two months after Thomas Jefferson was sworn in as President and her husband became his Secretary of State. Because Jefferson singled her out to serve on occasion as his official hostess, and because of her own charm and outgoing personality, she immediately became a social leader.

Thus, when at 41 she became First Lady, she was prepared for her brilliant eight year reign. Its splendor was only slightly dimmed in the summer of 1814 when she fled the President's house just ahead of the British who plundered and burned the mansion.

After the British left Washington, she and the President took up residence in Octagon House, a short distance from the blackened executive mansion, and there Dolley continued to entertain in the style which made her reign known as the golden age of social life in Washington.

After Madison's death in Virginia, she returned to Washington to reign as queen of society until her death on July 12, 1849, at the age of 81. In the annals of White House personalities she stands out as a truly unique and indomitable woman.

The bronze-dore mirrored plateau ordered by President Monroe in 1817, decorates the State Dining Room table in the 1960s.

James Monroe

chapter 5

A Return to Formality

Presidential entertaining took a turn toward increased formality when President James Monroe and his wife, Elizabeth Kortright Monroe, a New York aristocrat, moved into the Executive Mansion in December, 1817.

The Monroes had lived for ten years in Paris, where he was United States Minister to France. Elizabeth Monroe was accustomed to entertaining diplomats in the European fashion. This she much preferred, and as First Lady she decided to return to the graces and customs of the earlier Presidents, George Washington and John Adams.

It was in effect a new Executive Mansion into which the Monroes moved, for the structure had been rebuilt at a cost of $500,000 in the preceding three years since its burning by the British in 1814. Congress had appropriated $50,000 also for refurnishing.

This, however, was not nearly enough, and the East Room continued to be unused and unfurnished. To make the appropriation stretch as far as possible, Monroe sold his own collection of fine French pieces to the Government, and ordered more made to specifications for the mansion by Pierre-Antoine Bellange, the finest cabinet maker in Paris.

Mrs. James Monroe

Probably no piece they bought, however, has seen as much service as the beautiful mirrored plateau (surtout de table) selected by the Monroes as a centerpiece for their table. It has been used by every President since. Ordered from France at a cost of six thousand francs, (twelve hundred dollars), it was made in seven pieces and, extended its full length, is thirteen and a half feet long and two feet wide. Bronze garlands of fruit and vine decorate the rim where, at intervals, are figures of Bacchus and Bacchantes holding crystal vases for flowers and crowns for candles. Matching candelabra, fruit epergnes and vases were ordered at the same time.

Although silver-plated when purchased, it was gilded during the Van Buren administration and has remained that way ever since.

The state dinner service selected by the Monroes had five vignettes, representing strength, agriculture, commerce, art and science, set into a magenta border. In the center of the dinner plates is an eagle.

President Monroe left the decorating of the mansion to his wife, who chose light crimson satin for the upholstery of major pieces of furniture. Matching taffeta curtains were hung at the windows in the oval drawing room where Dolley Madison had used yellow damask.

It was in this setting that the Monroes held their first reception on New Year's Day, 1818. President Monroe, (a Virginian whom his father-in-law, Captain Lawrence Kortright, a former British Army officer, never felt was good enough for his daughter), received wearing old-fashioned knee breeches, silk hose and silver-buckle shoes of an earlier era. Men by this time had adopted the pantaloons or trousers President Jefferson had introduced in the Executive Mansion. But Monroe chose the old-fashioned styles for formal wear.

Mrs. Monroe, 48 and not in robust health, appeared

elegant, almost regal, wearing her hair piled in puffs on her head and adorned with ostrich plumes. She sat on a raised platform to receive the guests.

The weather was delightful, and the newly rebuilt mansion was thronged from twelve until three o'clock by senators, representatives, heads of government departments, foreign ministers, distinguished citizens, residents and strangers.

Refreshments were served, but Monroe, not being as wealthy as his predecessor, James Madison, served more coffee, tea, and a non-alcoholic punch and less wine.

Following are some of Mrs. Monroe's recipes:

A GOOD JULEP

One half pint of grape juice
One half pint of orange juice
One half cup of chopped mint
One pint of charged water
Juice of six lemons or limes

Mix fruit juice and mint and stand on ice one hour. Add sugar to taste. Add water and pour into glasses half filled with ice. A sprig of mint in each glass and serve at once.

LEMON CAKE

Mix together two cups of sugar; two cups of flour; two teaspoons of baking powder sifted into flour; one-half cup cold water; yolks of five eggs; whites of four eggs well beaten, pinch of salt. Bake in moderate oven.

Icing: Beat the white of one egg to a stiff froth, add powdered sugar until almost too stiff to stir, then add the grated rind and juice of one large lemon.

The formality of the Monroe reception was not too well received by Washingtonians who had been accustomed to the friendly warmth of Mrs. Madison's receptions and drawing rooms.

And Washington society had another cause to criticize Mrs. Monroe. She refused to make first calls as Dolley Madison did. She refused also to return calls. Instead, she relegated this chore to her elder daughter, Mrs. George Hay, who was often at odds socially with the ladies of the diplomatic corps.

Adding to the outrage was the fact that Washington

society was not invited to the wedding of the first Presidential daughter to be married in the White House. Only 42 guests were present for the wedding on March 9, 1820, of Maria Monroe, younger daughter of President and Mrs. Monroe, to her first cousin, Samuel Lawrence Gouveneur of New York.

President Monroe's ornate candelabra still decorates the mantlepiece in the East Room.

All this led to a virtual boycott of the Monroes' weekly drawing rooms during their first term in the President's House. In December, 1819, Mrs. William Seaton, wife of the editor of The National Intelligencer, wrote:

"The drawing room of the President was opened last night to a beggerly row of empty chairs. Only five females attended, three of whom were foreigners."

However, by the beginning of Monroe's second term, the resentment simmered down and the ladies decided they were missing more than they had thought by staying away. Mrs. Monroe's weekly receptions soon became popular.

The democratic crowds were described by one newspaper as follows: "In addition to the Secretaries, Senators, Foreign Ministers, consuls, auditors, accountants, and officers of the Army and Navy of every grade . . . farmers, merchants, parsons, priests, lawyers, judges, auctioneers and nothingarians—all with their wives and some with their gawky offspring crowd to the President's house every Wednesday evening; some in shoes, most in boots and many in spurs; some snuffing, others chewing and many longing for their cigars and whiskey punch at home."

Mrs. Monroe descended from her platform to receive guests at receptions and at their New Year's Day reception in 1821. This one attracted the largest crowd to attend a function at the Executive Mansion up to that time. She and the President and their two daughters stood near the door to greet guests as they entered.

Mrs. Monroe wore black velvet and pearls, looking very youthful with her cheeks discreetly rouged. Mrs. Hay wore crimson velvet, while Maria, the bride of less than a year, wore what may have been her wedding dress, a white satin trimmed in silver lace. Both daughters wore pearl necklaces and ostrich feathers in their hair.

Elizabeth Monroe added to the brilliance of her drawing rooms by spending $100 a night on wax candles alone, one contemporary account said. And wines were served on silver trays by servants in livery.

Author James Fenimore Cooper, a frequent guest of the Monroes wrote of one of Mrs. Monroe's drawing rooms as follows:

"On . . . Wednesday, Mrs. Monroe opened her doors to the world. No invitation was necessary, it being the usage of the wife of the President to receive company once a fortnight during the season without distinction of persons.

"We reached the White House at nine. The court (or grounds) were filled with carriages and the company was arriving in great numbers. On this occasion, two or three additional drawing rooms were opened though the frugality of Congress has prevented them from finishing the principal reception room of the building.

"I will acknowledge the same sort of surprise I felt at the castle garden fete at finding the assemblage so respectable in air, dress and deportment. The evening at the White House drawing room is in fact, a collection of all classes of people who choose to go to the trouble and expense of appearing in dresses, suited to an ordinary evening party. I am not sure that even dress is much regarded, for I saw a good many there in boots. The females were all neatly and properly attired though few were ornamented with jewelry. Of course, the poor and laboring classes would find little or no pleasure in such a scene.

"Squeezing through the crowd we achieved a passage to a part of the room where Mrs. Monroe was standing surrounded by a bevy of female friends. After making our bow here, we sought the President. The latter had posted himself at the top of the room where he remained all evening, shaking hands with all who approached. Near him stood all the secretaries and a great number of the most distinguished men. . . . "

Dinners given by the Monroes were elegant and formal. French cuisine was served at beautifully appointed tables, and the dinner hour was pushed back to first five o'clock and then six. Even though all Americans were not impressed with the formality of the Monroes' dinners, French and British visitors were.

PRESIDENT MONROE'S QUAKING JELLY

2 tablespoons gelatin (soak in 1/2 cup cold water for 5 minutes)
2 cups tomato juice
1/2 teaspoon chopped onion
1 tablespoon chopped parsley
1 small piece lemon rind
2 cups rich chicken stock

Bring to a boil all ingredients except gelatin. Remove from heat and strain. Dissolve gelatin in hot stock. Cool.

Flavor with:

2 tablespoons dry sherry
1 tablespoon lemon juice
salt and white pepper to taste

Pour into a wet mold. Chill until set. Before serving break lightly with a fork, serve in individual dishes garnished with sour cream and chopped chives. Serves six.

The wife of Monroe's Secretary of the Navy, wrote of one dinner at the White House:

"At Mrs. Monroe's we had the most stylish dinner I have been at. . . . The dishes were silver and set around the waiter. The plates were handsome china, the forks, silver and so heavy that I could hardly lift them to my mouth, dessert knives silver, and spoons very heavy—you would call them clumsy things."

Writing of dinner at the Monroes, James Fenimore Cooper said: "The dinner was served in the French style, a little Americanized. The dishes were handed around though some of the guests, appearing to prefer their own customs, coolly helped themselves to what they found at hand.

"Of attendants there were a good many. They were neatly dressed, out of livery, and sufficient. To conclude, the whole entertainment might have passed for a better sort of European dinner party at which the guests were too numerous for general or very agreeable discourse and some of them too new to be entirely at ease. Mrs. Monroe arose at the end of the dessert and withdrew, attended by two or three of the most gallant of the company. No sooner was his wife's back turned than the President reseated himself, inviting his guests to imitate the action. After allowing his guests sufficient time to renew, in a few glasses, the recollections of similar enjoyments of their own, he arose himself, giving the hint to his company that it was time to rejoin the ladies. In the drawing room coffee was served and every one left the house before nine."

A visit by Lafayette to this country at the end of 1824 provided a blaze of glory for the final months of the Monroe Administration. The French hero was honored at a state dinner at the White House, and was present at Monroe's last New Year's Day reception in 1825.

Although Monroe had his social squabbles, he was a popular President. And many of the social traditions established in his administration are still observed at the White House, such as the First Lady not making first calls, and not returning calls.

The charm and dignity of the colorful era of the Monroes have passed into legend. To follow them came the second Adams family, John Quincy and Louisa, who were to leave their own mark on Washington as social arbiters.

Before leaving the White House he repurchased from the Government his own furniture, and much of it may be seen today in the James Monroe Memorial Library at Fredericksburg, Virginia.

A foreign diplomat bound for a White House party during the John Quincy Adams Administration.

John Quincy Adams

chapter 6

PACE-SETTING PARTIES

Long before Mrs. John Quincy Adams, the London-born daughter of an American father and a British mother, became First Lady in 1825, she had proven herself an excellent hostess.

While her husband was Secretary of State, they had given one of the most brilliant parties Washington had witnessed up to that time. Given in January, 1824, it was in honor of the Hero of New Orleans, General Andrew Jackson, who had recently come to Washington as Senator from the state of Tennessee.

For the occasion, eight rooms of the Adams' big double house on "F" Street were stripped of superfluous furniture and decorated with tissue paper roses and evergreens. On the floor of the ballroom were chalk paintings of the American eagle, flags, and the words "Welcome to the Hero of New Orleans." The house was lighted with small glass variegated lamps almost everywhere, and decorated with artificial wreaths as well as the paper roses.

Mrs. Adams was dressed in a splendid gown of steel-colored llama cloth with steel-cut ornaments and a headdress of plumes. Standing at her side as she received the guests was the tall rugged Jackson. John Quincy Adams, who disliked parties and preferred going to bed early, was dressed severely in comparison to the glittering assemblage of diplomatic and military uniforms.

Some eight hundred guests attended, and many of them climbed on chairs and tables to get a view of the General as Mrs. Adams promenaded him through the rooms. At ten o'clock the doors to the dining room were opened upon a lavish table covered with refreshments of every description.

It was an event that Washington did not soon forget. Ironically, it helped advance the political ambitions of Jackson, who some four years later became John Quincy Adams' political rival.

The party was also a forerunner of what was to take place in the White House a year later when Adams was elected President. He was one of four candidates, including Andrew Jackson, Henry Clay, then speaker of the House, and William H. Crawford, Secretary of the Treasury. All failed to receive a majority of the electoral vote, and the choice was thrown to the House of Representatives, which selected Adams.

Louisa Catherine Adams, well-educated and highly accomplished, was qualified to preside as mistress of the White House, and she entertained lavishly there.

Because she and her husband were accustomed to the European style of entertaining from the many years he was in the diplomatic service before returning to become President Monroe's Secretary of State, they continued the formal style of entertaining established by the Monroes. Only they did so on a more lavish scale.

Mrs. Adams was also the style leader of the capital, her dress reflecting the Empire fashion then dominant in France.

Dinners at the White House during the Adams administration were described as "splendid but stiff." The food was continental in flavor because the Adams' had brought to the White House the French chef that had joined them during his diplomatic assignment in Amsterdam. Recipes for foods served by them included:

QUEEN CAKE

Take a pound of sugar, beat it fine with four yolks and two whites of eggs, half a pound of butter, a little rosewater, six spoonfuls of warm cream, a pound of currants and as much flour as will make it up. Mix them well together and put them in well-buttered patty-pans. Bake in a moderate oven for thirty minutes; take out and glaze (frost).

Others are:

NAPLES CAKE

Beat 12 eggs light, add to them one pound of flour, and one pound of powdered sugar. Continue to beat all together till perfectly light. Bake in long pans four inches wide, with divisions so that each cake when done will be four inches long and one and one-half inches wide.

TRIFLE

Put slices of Naples cake at the bottom of a deep dish, moisten it with white wine, and fill the dish nearly to the top with rich boiled custard; season half a pint of cream with white wine and sugar, whip it to a froth. As it rises, take it lightly off and lay it on the custard. Pile it up high and tastely. Decorate it with preserves of any kind, cut thin so as not to bear down on the froth by weight.

At their dinners in the White House, the table was set with a pink, white and gold Paris porcelain dinner set purchased by Adams when he was minister to Russia.

Louisa Adams played the piano and harp, and often the small private dinners they gave were followed by group singing to her accompaniment on these instruments.

President Adams, the son of the second president, found the weekly levees or receptions boring. Of one he wrote:

"This evening was the sixth drawing room. Very much crowded: 16 senators, perhaps 60 members of the

House of Representatives and multitudes of strangers among whom were the Institutors of Deaf and Dumb from Philadelphia, New York and Hartford. The heat was oppressive and these parties are becoming more and more insupportable to me."

An early 19th century writer described one of the Adams drawing rooms as follows:

"Gentlemen and ladies both attend, arrive about 8:00 and leave around 10:00. No lady goes by herself, but sometimes one gentleman waits on two but most commonly only one. They mostly all stand, or move around the rooms; the ladies always resting on the left arm of the gentlemen, when he attends to only one. The guests, on arrival, find the President and his Lady standing in the upper part of the room, but at some distance apart with some of their particular friends at their side. The guests all make their way through the crowd, toward the President and salute him; and those of his acquaintance shake hands with him, and then pass on to salute his lady.

"There is no established rule, or formal regulation for the support of order, more than that which prevails in other assemblages of fashionable and genteel people in other places of the city.

"There are three large furnished rooms, besides the two ante-chambers, the middle one is round and is called the salon, where the President and his Lady receive their guests; they are generally so full that it is with much delay and edging difficulty that a gentleman can get through to the upper end of the salon to salute the President.

"The company is treated with coffee, tea, and a variety of cakes, jellies, ice-cream and white and red wine, mixed and unmixed, and sometimes other cordials and liquors, and frequently with West India fruit; all of which are carried about the room amongst the guests upon large trays by servants dressed in livery; each one takes from it what he pleases, when an opportunity offers, which at some of the fullest levees, may not happen very often not because there is any scarcity of refreshment, but the difficulty the waiters have in making their way through the crowd with their trays. . . . After some part of the company have retired, so as to give more room for the waiters to move freely about the rooms with refreshments, everyone is furnished bountifully.

"At the President's levees there is commonly no other amusement but conversation so that those who happen to be quite strangers, are merely spectators; and at every levee there are many of that class so that the novelty, or the presumed honor of being one of a part of the highest order in the United States are all the inducements that an entire stranger has to go there."

The following are some recipes from a cookbook owned by Mrs. Adams:

SEED CAKES

Take a pound of fine sugar sifted and a pound of the best butter; beat it for half an hour; add eight eggs and a pound of flour; beat all together and add a pennyworth of carraway seeds.

APPLE FRITTERS

Take the yolks of six eggs, and the whites of three. Beat them well together and put to them a pint of cream or milk. Add four of five spoonfuls of flour and a glass of brandy, half a nutmeg grated, and a little ginger and salt. Your batter must be pretty thick. Then slice apples in rounds, dip each round in butter and fry in good lard (shortening) over a quick fire.

FRIED OYSTERS

Make a batter of milk, eggs and flour. Wash and dry oysters. Dip them in the batter, then roll them in bread crumbs and a little mace beat fine. Fry in hot butter or lard.

Sometimes the crowds were so large the guests seemed to be lifted from their feet and carried along without any effort of their own. At one especially crowded evening reception, a general was robbed of his wallet containing eight hundred dollars.

Mrs. Adams opened her doors to men of every political opinion and every station. She had three grown sons who often found the throngs at the receptions dull company. One son referred to the guests at one drawing room as ragamuffins.

Another son, John, who served as his father's private secretary, insulted some of the guests one evening. Perley Poore, a 19th century reporter, told the story of a Washington newspaper editor, Russell Jarvis, who had supported Jackson for the presidency, going to an Adams reception accompanied by his wife and a party of visiting relatives from Boston.

They were received courteously by the President and Mrs. Adams and passed on to an adjoining room. There they stood opposite John Adams who was talking with a minister. The latter asked "who is that lady?" to which John replied in a tone loud enough to be heard by all around: "That is the wife of one Russell Jarvis, and if he knew how contemptibly he is viewed in this house, they would not be here."

Jarvis and his party left immediately and he wrote a letter to the President's son demanding an apology. The latter replied he had no apology and wished no correspondence with him.

A few days later John Adams went to the Capitol to deliver messages from the President to each House of Congress. As he was walking through the rotunda, from one house to the other, he met Jarvis who, the story goes, pulled his nose and slapped his face. A scuffle ensued, until the two were parted by a Congressman from Maryland. The President made a formal complaint to Congress and the House appointed a select committee to investigate, but it did not recommend any punishment in the case.

Dancing at the White House began in the final years of the Adams administration and became a popular form of entertaining at the parties given by the Adams sons for their young friends.

John, the Adamses' second son, was married in the White House, the only Presidential son in history to have that distinction. His bride was his coquettish cousin, Mary Catherine Hellen, who lived with the Adamses' and had been romantically involved with John's two brothers, George Washington, and Charles Francis, before becoming betrothed to him.

The wedding was held on February 25, 1828, on Monday evening, in the oval drawing room furnished with the elegant French furniture that Monroe had ordered from France. Neither George, who was in Boston, nor Charles at Harvard, attended the ceremony, performed by the Reverend William Hawley of St. John's Church on Lafayette Square.

Garlands of ribbons and flowers, hung by the bridesmaids, decorated the room. A wedding supper was served afterwards. Three evenings later a reception honoring the newly married couple was held in the yellow drawing room, (now the Red Room) and the President joined in dancing the Virginia Reel.

On December 2, 1828, a daughter was born to John and Mary in the White House. She was the first girl child to be born in the mansion and the first child to be christened there. Her christening was one of the last social functions of her grandfather's administration and was attended by members of the Cabinet, officials of Government and other distinguished guests.

The most brilliant social event held by President and Mrs. John Quincy Adams in the White House was a dance in the still unfurnished East Room to celebrate their departure from the mansion.

It was held in late December and the Marine Band played the latest, liveliest tunes for dancing. The ladies wore their most elegant gowns and the officers came in dress uniforms. Louisa broke the custom of receiving the guests surrounded by Cabinet wives, and passed from group to group in welcoming her guests. The President, for once enjoying a social event, stood at one end of the room shaking hands and taking bows.

It was an event that rivaled in brilliance the party Louisa Catherine had given five years earlier in honor of Andrew Jackson, who would succeed them in the White House.

Andrew Jackson

chapter 7

The Hospitality of Old Hickory

Hospitality was free and lavish at the White House during the eight years Andrew Jackson, the soldier-frontiersman, was President. Although it lacked the polish and formality of the Adams and Monroe administrations, it was never dull.

Twenty thousand admirers of the "Hero of New Orleans" thronged the mansion at a reception following Jackson's inauguration on March 4, 1829. They all but demolished the White House and barely missed crushing him.

Never before, nor since, has the historic mansion seen such an unruly, enthusiastic hero-worshipping mob. After his inauguration at the Capitol, Jackson rode on horseback to the White House followed by streams of people. He arrived to find the grounds and house already thronged by thousands more.

From all over the country they had come—hunters from Kentucky, Indian fighters from Tennessee, frontiersmen from the Northwest, as well as planters, squires and city folks from Virginia, Maryland and the Atlantic seaboard.

When Jackson entered the mansion, the crowd surged in behind him. Men in rough muddy boots and women in their finest brocades and satins stood on the chairs and sofas for a look at their hero. To lure some of the horde outside, buckets of punch were placed on the lawn, but it didn't hold the crushing crowd outside. Back they came, pressing to get closer to "Old Hickory." He retreated slowly and smiling with his face to the surging crowd. Finally his back was to the wall and he could go no further. Some of his friends linked arms to form a barrier around him, and succeeded in spiriting him out a back door.

As the crowd surged, windows were broken out and some jumped out onto the grounds. In the melee much White House china and crystal were smashed; rugs and furniture were damaged and broken. It was a costly but triumphant beginning for the popular President, who still mourned the passing, three months earlier, of his beloved wife, Rachel Donelson Robards.

She had died at 61 from a stroke after having overheard malicious gossip about her fitness to serve as First Lady. She was shopping for a gown to wear to her husband's inauguration when she overheard the remarks, and the ball gown became her shroud.

Although Jackson had little heart for entertaining, he recognized the social responsibilities of the presidency and asked Emily Donelson, his wife's attractive, auburn-haired niece from Tennessee, to become his official hostess.

Emily and her husband, Andrew Jackson Donelson, Rachel's nephew, whom the Jacksons had adopted as their own son, moved into the White House with "Old Hickory" and tried to make it home. It was here that Emily gave birth to three of her four children.

The christening of one of her daughters, Mary, who was born in the upper western chamber facing Pennsylvania Avenue, was a celebrated event in the White House.

 The East Room, which Jackson had decorated with

$10,000 provided by Congress, was softly lighted by many tapers gleaming from three handsome chandeliers. The walls of fine lemon-colored paper reflected the light, and provided a brilliant background for the handsome mantels of black marble with Italian black and gold fronts that framed the fireplaces. Above each mantel was a huge mirror.

Cora Livingston, a belle of the time whose father later became Jackson's Secretary of State, served as godmother to the baby, and the President shared honors as godfather with Martin Van Buren, his first Secretary of State. Jackson held the baby tenderly during the christening, pleased that the little lady had deserted Van Buren for him. Members of the diplomatic corps, the Judiciary and Congress were invited. Among the guests was Robert E. Lee, then a young lieutenant, and the girl who became his bride, Mary Custis.

The first levee during the Jackson administration was held on Thursday, January 10, 1830, and was described at the time as a "splendid one." It began at 5 o'clock in the afternoon and for five hours the President and his cabinet stood shaking hands with friends and foes alike.

Emily Donelson and Mary Eastin, a niece who lived at the White House, also welcomed the guests. Both women were dressed in American calico without ruffles or ornaments, showing their loyalty to the Jackson party. During the campaign that elected Jackson, lady partisans had distinguished themselves by dresses of American calico rather than the French calico worn by persons of social status.

At parties during the Jackson administration, the younger set sought the dance floor while the older guests gathered around the bowl to discuss questions of the day. The punch bowl was usually filled with a concoction called "Daniel Webster Punch." It was composed of Medford rum, brandy, champagne, arrack (a distillation of rum from the Far East), maraschino, strong tea, lemon juice and sugar, and was a strong mixture with a kick.

Jackson held two dinners a week at the White House and an assembly once a week between Christmas Day and Ash Wednesday, to which all the respectable ladies of the city who danced were invited.

His dinners were events to remember. He employed a French chef and imported servants from his Tennessee home to provide gastronomic delights for his guests. He himself often provided the entertainment. At one of his dinners he sang "Auld Lang Syne", much to the merriment of his guests.

John R. Montgomery, a Lancaster, Pennsylvania, lawyer who attended one of Jackson's dinners in February, 1834, gave this account of it:

"About one-half after six o'clock we sat down to dinner. The table was very splendidly laid and illuminated. There was a large chandelier hanging over the middle of it with 32 candles besides those on the table, mantels and on the piers. The first course was soup in the French style; then beef bouille, next wild turkey boned and dressed with brains; after that fish, then chicken cold and dressed with white, interlaid with slices of tongue and garnished with dressed salled; then canvas back ducks and celery; afterwards partridges with sweet breads and last pheasants and old Virginia ham. The dishes were placed in succession on the table, so as to give full effect to the appearance and then removed and carved on a side table by the servants. The first dessert was jelly and small tarts in the Turkish style, then blanche mode and kisses with dryed fruits in them. Then preserves of various kinds, after them ice cream and lastly grapes and oranges.

"The wines on the table were Sherry and Port to drink with soups and the first course of meats. When the wild turkey and fish were served, Madeira was handed and while the wild fowl was eaten, champagne was constantly poured out by the servants; after these were gone through with, Claret was substituted to be taken with the dessert and old Cherry was put on to drink with the fruits. As soon as all had taken what their appetites could possibly endure, we left the table and returned to the drawing room.

" . . . French servants know how to guard American appetites from the ill effects of too much indulgence and in helping to any one dish do it so sparingly as to leave room for another and yet another to follow. We were at the table until nearly 9 o'clock and were eating and talking all the time. The President was very affable and his niece, Mrs. Donelson, the lady of the house, near whom I sat, a very agreeable woman, so that the evening passed very pleasantly. Soon after our return to the drawing room, a cup of coffee was handed, then the ladies played the piano and sung and after this a glass of 'liquer' was sent around as the signal for adjournment and the party broke up about one-half past nine o'clock."

Jackson's state dinners had a much greater degree of richness, formality and dignity. Mary Coffee, one of the young ladies who lived at the White House for a few months, described them as "stiff and ceremonious as possible, so they are not remarkably agreeable."

Jackson never liked to be alone, and that is why he had the mansion bulging at the seams with a near-overflow of relatives and friends sharing his home. He also loved having his grandchildren with him and spent many hours laughing and talking with them at play.

One of the most heartwarming parties he gave during his presidency was the Christmas party for children in 1835. There were six little ones around his own hearthside—the first time there had been so many with him at Christmas. The invitations went out six days before Christmas to about one hundred little ones. It read:

"The Children of President Jackson's family request you to join them on Christmas Day, at four o'clock, P.M., in a frolic in the East Room."

Children of the Jackson household wore their Sunday best to the party. Six-year-old Mary Donelson wore a blue cashmere dress; three-year-old Rachel Jackson a pink one with kid slippers; John Donelson wore a Highland plaid suit, and his brother, Jackson, a little suit with brass buttons.

As the young guests arrived they were greeted by the President, Emily Donelson, Sarah Jackson (who had married the President's adopted son, Andrew), and the children of the White House. They went directly to the East Room where their childish shyness disappeared as the games began: Blind Man's Buff, Hide and Seek, Puss in the Corner, and forfeit games. One little boy incurred a penalty and was sentenced to stand in the middle of the room on one leg and say:

"Here I stand all ragged and dirty,
If you don't come and kiss me I'll run like a turkey."

No one volunteered to kiss him and he strutted around the room like a turkey gobbler much to the amusement of the children and the grownups.

The President himself led the way as the Marine Band struck up a march, and the little guests followed him into the State Dining Room which was gaily decorated. In the center of the maltese-cross shaped table was a towering pyramid of snowballs decorated with colored icicles and topped by a gold game cock with wings outspread. At one end of the table were dishes of frozen fruits and at the other, candies, cakes and pastries in fancy designs.

The President's levee.

After supper the snowballs in the center of the table were handed out and the youngsters had a snowball battle in the East Room. The balls were made of non-combustible cotton coated with starch and in the center of each was a candy kiss. As the balls struck they broke, sending a shower of snow on the little heads.

When it came time to go the little guests bade the friendly President good night as he stooped down to shake each little hand. It had indeed been a Merry Christmas for the President.

The night before he had given a party for grown-ups, standing in what is now the Blue Room to shake hands with three hundred invited guests. Ices, wine and lemonade were passed as the guests promenaded and talked in the East Room. At 11 o'clock, a supper was served in the dining room and one guest, describing it, said "It surpassed everything of the kind I ever saw before . . . we had everything. This party could not have cost the President much short of $1,500!"

At heart President Jackson was a matchmaker who delighted in keeping up with the romances of his nieces and nephews and their friends. He gave White House weddings to two brides and welcomed a third to the mansion during his eight years in the presidency.

His adopted son, Andrew Jackson, Jr., was married in Philadelphia on November 24, 1831, to Sarah Yorke, and brought her to the White House to live.

To introduce the beautiful young bride who filled a daughter's place in his life, Jackson gave a dinner for about forty persons including the Cabinet and the Diplomatic Corps. He waived protocol, (over which there was then a dispute as to whether the whole cabinet or the diplomatic corps should rank next to the Secretary of State), and himself escorted the bride into the dining room, asking others to follow at will. Some three hundred additional guests were invited for music and dancing following the dinner.

Four months later, Jackson was again involved in a love match. This time he gave a White House wedding for Mary Eastin, the granddaughter of Rachel's brother, Colonel John Donelson. She was married to Lucius Polk, a second cousin to the future President James K. Polk, in a brilliant East Room ceremony.

Before the year was out there was another wedding in the same setting. This time it was a Catholic ceremony—the first in the White House—uniting Mary Anne Lewis, daughter of Jackson's old friend who had been his quartermaster at New Orleans, Major William Berkeley Lewis, to Alphonse Pageot, a secretary at the French Legation.

The 18-year-old bride, who was living in the White House, met her future husband at the Christmas party given by the President the preceding year.

Guests feast on a 1400 pound cheese presented to President Jackson.

Jackson's administration had been a lively one, socially speaking, and it ended with one of the most talked-about parties in the history of the mansion. That was the reception on Washington's Birthday, 1837, to which the public was invited. He had received a fourteen hundred pound cheese as a gift and decided to share it with the public on this occasion.

It was placed on a stand in the hall beneath a huge likeness of the President which hung on the wall surrounded by an eagle and stars. With knives, forks, spoons and their fingers, the guests picked away at the cheese as they pushed their way to it. Two hours later, all that was left were the crumbs ground by many footsteps into the floor and carpets, and the fragrance that lasted many weeks.

While the rank and file public whittled away at the cheese, diplomats in their gold-braid coats and officers in uniform and their ladies mingled in the East Room.

Jackson, not well, sat in a chair in the oval room for nearly three hours and received his friends with his usual formal courtesy. By his side stood Mrs. Andrew Jackson, Jr., now acting first lady, because Emily Donelson had returned to Tennessee where she died the preceding winter.

Before the crowd dispersed, Jackson grew weary and went upstairs to rest. His place was taken by the President-elect, Martin Van Buren, who received the multitude until the last guest departed just before 4 o'clock.

Every President had his social problems in the White House but none more than did Jackson. His problem stemmed from the fact that the wives of high government officials, and even his own official hostess, Emily Donelson, refused to receive pretty Peggy O'Neale Eaton, the daughter of an innkeeper in whose establishment Jackson had lodged before moving to the White House.

She was the wife of Jackson's Secretary of War, but because of her bad reputation and the gossip that she married him without getting a divorce from her first husband, other cabinet wives refused to recognize her socially.

The President, remembering the malicious gossip about his own beloved wife, defended Peggy Eaton and put pressure on members of his cabinet to force their wives to recognize her. All but one of the cabinet members resigned instead.

The problem was finally solved when Peggy's husband was made American minister to Spain and he took her with him. She became a favorite in Spanish court circles and never lost her quest for excitement. Years later she eloped with her granddaughter's music teacher.

White House table decorations have always been of interest. This recent table setting shows a low arrangement of flowers.

Here is a recipe for one of the desserts Jackson served in the White House and at his home in Nashville, Tennessee, "The Hermitage."

BURNT CREAM
(Creme Brulèe)

2 cups cream
4 egg yolks
2½ tablespoons granulated sugar
¼ cup light brown sugar
1 teaspoon vanilla

Heat cream in double boiler. Beat egg yolks, adding granulated sugar gradually. Remove cream from heat and pour over egg mixture very slowly. Add vanilla. Pour into 1½ quart baking dish or mold. Place pan in hot water and bake uncovered at 325 degrees about 45 minutes or until set. When custard is set, sprinkle with sifted brown sugar and place under broiler until sugar melts. Chill and serve very cold. Serve with sliced strawberries or other fruits if desired. Serves six.

Martin Van Buren

chapter 8

Years of Contrast

Social life at the White House proceeded at a sedate pace with an occasional outburst of brilliance during the next 24 years, while eight presidents held office successively.

Suave, diplomatic Martin Van Buren had been a widower for ninteen years when he was elected to succeed General Andrew Jackson to the presidency in 1837. Although he and Jackson had been good friends, and Jackson participated in Van Buren's inauguration, as soon as the latter got into the White House he changed the social patterns of his popular predecessor.

Perhaps one reason was the disorderly crowd that flocked to the Executive Mansion following Van Buren's inauguration. For three hours a surging tide of humanity swept past the new President who received them in the Blue Room. It was a reception as disorderly as could be imagined. At four o'clock in the afternoon, members of the Diplomatic Corps called on the new President and presented their congratulations. In his response, Van Buren, weary from his exertions, addressed them as the "Democratic Corps."

After that he banned the public from receptions at the White House except for one formal levee on New Year's Day. On this occasion no refreshments were served, and his guests had to be content to listen to the Marine Band, and discuss the weather with each other.

This does not mean Van Buren was a cold or unfriendly person. He had refined manners and entertained frequently at small, elegant dinners calling on various "grand dames" in the capital to serve as his hostess. Among them was Dolley Madison, still in her prime.

He also discarded the worn furniture of Jackson's era—auctioning part of it off—and spent $25,000 to replace it. He had the furniture in the oval drawing room reupholstered in blue instead of the crimson of earlier years, and the name of the room was changed to the Blue Room. He ordered tinted chinaware, gold-plated flatware, and emerald green finger bowls.

In addition to his own exclusive parties at the White House, he urged members of his Cabinet to give a dinner each season and he attended them.

Such lavish entertaining for an exclusive few did not set well with the mood of the people who were then beginning to experience the first great American depression, and Van Buren quickly became unpopular. His costly decorations of the White House, and lavish entertaining of his friends while the public went without so much as a cup of coffee at his annual levee became a campaign issue.

Silver water pitcher inscribed "Martin Van Buren."

A favorite soup during President Van Buren's years in the White House was Boula.

BOULA

1 can green turtle soup	pepper
1 can green pea soup	whipped cream, slightly salted
1/2 cup Sherry	Parmesan cheese
1 tablespoon butter	

Mix soups and bring to a boil. Add sherry and freshly ground pepper to season. Fill individual bowls and top each with a spoonful of whipped cream. Sprinkle with cheese. Place under broiler to brown topping slightly and serve immediately. Serves four.

When his son married the beautiful Angelica Singleton, of South Carolina, and brought her to the White House to serve as his father's official hostess, it didn't help Van Buren's popularity. The crowds turned out at her first appearance at a levee on January 1, 1839. She was easy and graceful in her manners, free and vivacious in her conversation and is said to have borne the fatigue of the three hour levee with a patience and pleasantry which must have been a trial.

However, she received the guests from an improvised throne on which she sat in a purple velvet dress with a long train. Three ostrich plumes were in her dark hair, and a pendant held by a rope of pearls decorated her forehead. This smacked of royalty, and indeed, Angelica did have royal ideas, for she had returned recently from a delayed honeymoon to Europe where she and her husband were presented at the courts and entertained royally. She brought some of the aura of her royal treatment back with her.

At the levee, the guests stared at her with avid curiosity and then passed on to gossip that her appearance was undemocratic and her costume an example of the President's lavish personal living standards for which the country must pay.

Later that year they voted Van Buren out of the White House and selected as his successor a man whose campaign emblem had been a log cabin.

That man was William Henry Harrison, a Virginian by birth and education, who had migrated to Ohio in early manhood, and who won the presidency with the log cabin symbol. And he proved himself to be a man of the people.

On a raw inauguration day he rode without top-coat to the Capitol to take the oath of office, and spoke for one and one-half hours out of doors to the crowd that gathered to cheer him. Then he returned to the White House to stand in the reception room and greet the thousands who filed by to shake his hand. They entered at the North Portico and then passed down the rear steps and out through the garden on the South.

William Harrison

Harrison's wife, Anna, did not accompany him to Washington for she was ill at their home in North Bend, Ohio. Accompanying him to serve as his hostesses were a widowed daughter-in-law, Jane Findlay Irwin Harrison, and the latter's aunt, Mrs. James Findlay.

They had little time for entertaining, however. Not long after taking office, Harrison took his market basket early one morning and walked to market without wearing a topcoat. It was a cold, stormy day and he was overtaken by a shower and got wet. He refused to change his clothes when he returned to the White House and came down with pneumonia. He did not recover, and one month from the day he took office, he died.

John Tyler, the courtly Virginian who had been elected vice president with Harrison under the slogan "Tippecanoe and Tyler too," succeeded to the presidency, and during the next three years there were occasional periods of brisk activity in White House entertaining.

The Tylers entertained in the Virginia manner and the style was warm and lavish. However, Mrs. Tyler, the former Letitia Christian of New Kent County, Virginia, was in ill health and rarely appeared at social functions. One of her rare appearances was at the wedding on January 31, 1842, of her daughter, Elizabeth, to William Waller, of Williamsburg, Virginia, in the East Room.

The wedding guests consisted of cabinet members, foreign ministers and personal friends, including Dolley Madison who was still a queen of Washington society.

Because of Mrs. Tyler's ill health, her daughter-in-law, Mrs. Robert Tyler, the former Priscilla Cooper, (daughter of the actor Thomas A. Cooper), served as official White House hostess while Tyler's older daughter, Letitia, took care of her mother.

Being official hostess at the White House and a young mother at the same time sometimes posed a hardship for the beautiful Priscilla Tyler. She wrote:

"My first state dinner is over; oh! such a long one, our first dinner in the State Dining Room. I was the only lady at the table . . . I tried to be cheerful as possible, though I felt miserable all the time as my baby was crying, and I received message after message to come to the nursery."

At another state dinner, Priscilla, pregnant with her second child, and tired from the strain of caring for her four-months-old daughter, Mary Fairlie, grew pale as the dessert was being served and fell back from the table in a faint. Daniel Webster gathered her in his arms and carried her from the table, but before he got away, Robert Tyler poured a pitcher of ice water on the two of them. Priscilla had to be taken to her room, and Mr. Webster shaken off, dried and brushed before he could resume the dinner.

President Tyler held two dinners a week while Congress was in session. One was usually confined to twenty guests—all of them men—from different parts of the country visiting Washington, and the second, for forty guests, both ladies and gentlemen, included officials of various government departments, and members of the diplomatic corps and their wives.

Desserts served during this period were:

POUND CAKE

Wash the salt from a pound of butter and rub it until it is soft. Have ready a pound of flour sifted, one pound of powdered sugar, and twelve eggs well beaten; put alternately into the butter, sugar, flour, and the froth from the eggs—continuing to beat them together till all the ingredients are in and the cake is light; add some grated lemon peel, a nutmeg, and a gill of brandy

Butter the pans, fill and bake to a golden brown. This cake makes an excellent pudding if boiled in a large mold and eaten with sugar and wine. It is also excellent when boiled and served with melted butter, sugar and wine.

LITTLE PLUM CAKES

Prepare batter as directed for pound cake. Add raisins and currants, bake in small tin shapes and ice when done.

MACAROONS

Blanch a pound of sweet almonds, pound them in a mortar with rosewater; whip the whites of seven eggs to a stiff froth, put in a pound of powdered sugar, beat it some time, then put in almonds. Mix them well and drop on sheets of paper buttered. Sift sugar over and bake them quickly. Be careful not to let them get discolored.

The drawing room was open every evening informally until ten o'clock, when the family rose and retired and the doors were closed. Occasionally during the winter months, a private ball with dancing was held with special invitations going to officials and friends. During mild weather a concert by the Marine Band or some other musical group was presented on the South side of the mansion for the recreation of the public; and once a month, a public levee was held in addition to the general receptions on January 1st and July 4th.

When the Prince de Joinville, son of Louis Philippe of France, visited Washington in 1841, President Tyler gave an official dinner and a ball in his honor. Daniel Webster, as Secretary of State, was there with his wife, as were other members of the cabinet, members of the diplomatic corps in full court dress, and army and navy officers in dress uniform.

Elizabeth Tyler, the President's pretty teen-aged daughter and the 23-year-old tall, handsome Prince led a quadrille d'honneur in the East Room, after which dancing became general.

Some members of the cabinet had advised against dancing as undignified, but the President, who was fond of the Virginia Reel, felt a young prince would enjoy this more than a more formal evening.

In 1842 both Charles Dickens and Washington Irving attended a public levee given by President Tyler. Dickens wrote this historic account of the event:

"I went with my wife about ten. There was a pretty dense crowd of carriages and people in the courtyard and so far as I could make out, there were no very clear regulations for the taking up or setting down of company. There were certainly no policemen to soothe startled horses, either by sawing at their bridles or flourishing gruncheons in their eyes. But there was no confusion or disorder. Our carriage reached the porch in its turn, without any blustering, shouting, backing or other disturbance; and we dismounted with as much ease and comfort as though we had been escorted by the whole metropolitan force from A to Z inclusive.

"The suite of rooms on the ground floor were lighted up, and a military band was playing in the hall. In the smaller drawing room, the center of a circle of company were the President and his daughter-in-law, who acted as lady of the mansion, and a very interesting, graceful and accomplished lady too. One gentleman who stood among the group appeared to take upon himself the functions of a master of ceremonies. I saw no other officers or attendants and none were needed.

John Tyler

"The great drawing room and the other chambers on the ground floor were crowded to excess. The company was not in our sense of the term select for it comprehended persons of very many grades and classes; nor was there any great display of costly attire; indeed, some of the costumes may have been for aught I know, grotesque enough. But the decorum and propriety of behaviors which prevailed were unbroken by any rude or disagreeable incident; and every man, even among the miscellaneous crowd in the hall who were admitted without any orders or ticket to look on, appeared to feel that he was a part of the institution and was responsible for its preserving a becoming character, and appearing to the best advantage.

"That these visitors, too, whatever their station, were not without some refinement of taste and appreciation of intellectual gifts and gratitude to those men, who by the peaceful exercises of great abilities shed new charms and associations upon the homes of their countrymen and elevate their character in other lands, was most earnestly testified by their reception of Washington Irving, my dear friend, who had recently been appointed Minister at the Court of Spain and who was among them that night in his new character, for the first and last time before going abroad.

"I sincerely believe that in all the madness of American politics, few men would have been so earnestly, devotedly and affectionately caressed as this most charming writer; and I have seldom respected a public assembly more than I did this eager throng."

Mary Fairlie Tyler, President Tyler's granddaughter, dressed as a fairy princess for a children's costume party at the White House.

Another happy event was the masquerade party given to celebrate the birthday of the President's granddaughter, Mary Fairlie Tyler. Dressed as a fairy with a diamond star on her forehead and in her hand a silver wand, she represented Queen Titania. She welcomed the young elite of Washington at the door to the East Room, and among the guests was Dolley Madison's little grandniece, Adele Cutts, who came as a flower girl. Students from a fashionable dancing school in the capital put on a program as part of the entertainment.

At his dinners President Tyler was a gracious host who did everything within his power to put his guests at ease. This determination is reflected in a story handed down from generation to generation to his descendants. As the story goes, he invited a senator to one of his state dinners who was unaccustomed to formal entertaining. When the fingerbowls were placed before the guests, this senator picked up his bowl and drank the water. Not wishing the senator to be embarrassed by his mistake, President Tyler drank the water from his fingerbowl, and taking their cue from him, all the others around the table did likewise.

In September 1842, the invalid Mrs. Tyler died and the White House was plunged in mourning. Except for a few functions, entertaining at the White House was discontinued that winter. However, select friends were invited as personal guests of the President and his family for games of whist or evenings of music. Among these was the attractive twenty-three-year-old Julia Gardiner, daughter of a former New York state senator who was spending the social season in Washington with her parents.

Julia had met President Tyler the preceding January at a White House reception. At that meeting she was impressed by the "silvery sweetness" of the President's voice, his "incomparable grace of bearing" and the "elegant ease of his conversation."

Soon after he became a widower, Tyler began to take a romantic interest in Julia. They courted in the White House and before long he was asking her again and again to marry him. Her answer was "No, no, no." But a personal tragedy was to play a part in her decision. For when Julia and her father were members of a distinguished party of three hundred who accompanied the President on an excursion down the Potomac River aboard the Navy's new frigate, the "Princeton" one of the guns being fired experimentally, exploded, killing her father. Some time later Julia relented, and accepted Tyler's proposal. So on June 26, 1844, they were married in a secret ceremony in New York. Two days later they gave a joyful wedding reception in the flower-filled White House—the first President to celebrate his own wedding there.

The reception was held in the East Room where Tyler had begun his courtship with his new bride. John C. Calhoun, Tyler's Secretary of State since the "Princeton" catastrophe, escorted the bride to the supper table and cut the wedding cake for her. The new Mrs. Tyler wrote of the event: "The company who waited upon me with their most respective compliments comprised the talent and the highest station in the land. For two hours I remained upon my feet receiving quite in a queenlike state."

The White House during the Tyler Administration.

President Tyler enjoyed eating duck and had it for breakfast the morning following his wedding.

ROASTED CANVASBACK DUCK

Procure a fine fat canvasback duck (picked, singed and drawn). Wipe it and throw inside a light pinch of salt; run in the head from the end of the head to the back, truss it and place in a roasting pan. Sprinkle a little salt on top and cook for eighteen minutes in a brisk oven. Place on a very hot dish, untruss and throw in two tablespoonfuls of white broth. Serve with six slices of fried hominy and currant jelly for garnish.

As First Lady, Julia Gardiner Tyler brought a brillance to White House entertaining that rivaled the golden era of Dolley Madison. However, the public looked upon it with some misgivings.

She held drawing rooms at which she received while seated in a large armchair on a slightly raised platform, surrounded by twelve maids of honor dressed alike. Julia herself wore purple velvet with a train and a headdress of bugles resembling a crown. The ostentation was a little too much for some of the capital's elite, who shook their heads in disapproval.

However, Julia loved it and went on her lavish way. She initiated the tradition of having the Marine Band play "Hail to the Chief" and won her husband's approval for dancing the polka and waltz in the White House.

Writing of one of her social successes, Julia said, "Last evening I had a most brilliant reception. The British Minister, Pakenham, was there with his Secretary, and devoted to me. At least fifty members of Congress paid their respects to me, and all at one time. I did not enter the room until they were assembled. It really presented an array and it was imposing to see them all brought forward and introduced one by one."

The most brilliant event of her "reign," however, was the final ball. Over two hundred invitations were issued to the social and political elite from up and down the Eastern seaboard. As the guests arrived, their names were announced and they then passed on after exchanging congratulations with the President, his bride and ladies of his household who stood in a line at the side of the room.

Julia was dressed in embroidered satin, partly covered with looped lace, and wore a Shepherd bonnet with ostrich feathers and diamond ornaments. The carpets in the East Room had been removed and the floor waxed for dancing. In the recess of the great windows, draped with blue damask and covered with an American flag, was the orchestra. Julia opened the ball by dancing with the Secretary of War, and then there followed general dancing. The ladies were attired with an abundance of glittering jewels unequalled on any other occasion up to that time, according to eyewitness accounts. The supper was magnificent and wine flowed like water. It was rumored that the event cost the President nearly two thousand dollars.

Julia had her chance to shine once more before leaving the mansion with her husband. It was at the dinner party she gave for the new occupants of the White House, President-elect and Mrs. James K. Polk.

On this occasion she wore on a gold chain around her neck the golden pen with which President Tyler had signed the treaty bringing Texas into the Union—the crowning achievement of his presidency.

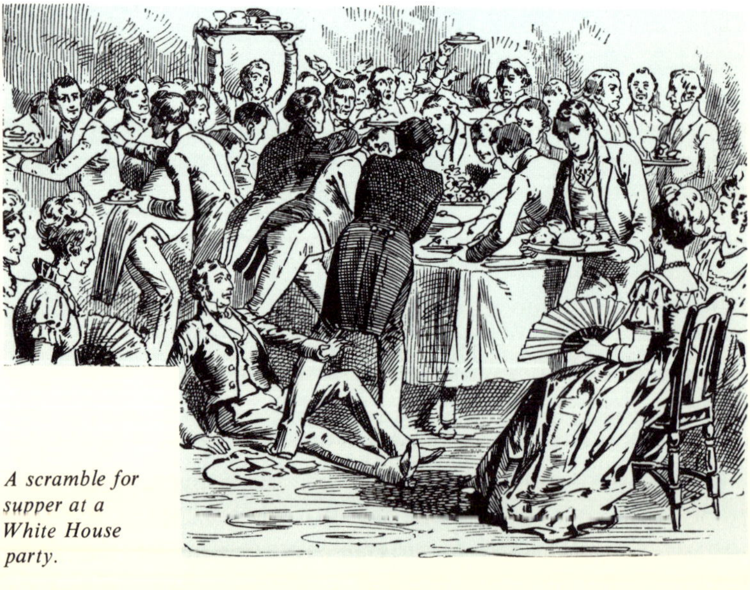

A scramble for supper at a White House party.

James K. Polk

chapter 9

More Contrasts

Dancing, card playing and wine drinking were banned from the White House during the James K. Polk administration. Sarah Childress Polk, who was 42 at its start, had a Spanish cast to her beauty and was charming and clever in conversation. She was a popular First Lady despite the limitations and restrictions she imposed on entertaining in the White House.

She held receptions twice a week on Tuesday and Friday evenings but the guests were offered no entertainment after being received by the Presidential party, except to promenade in the East Room and exchange the gossip of the day. However, there were few complaints.

Henry Clay was a guest at many of the Polk dinners which were subdued affairs where the conversation tended to outshine the food. At one he remarked to Mrs. Polk: "Madame, I must say that in all my travels wherever I have been in all companies and among all parties, I have heard but one opinion of you. All agree in commending in the highest terms your excellent administration of the domestic affairs of the White House. But as for that young gentleman there," he continued referring to the President, "I cannot say as much. There is some little difference of opinion in regard to the policy of his course."

Mrs. Polk graciously replied: "I am glad to hear that my administration is popular. And in return for your compliment, I will say that if the country should elect a Whig next fall, I know of no one whose elevation would please me more than that of Henry Clay. And I will assure you of one thing. If you do have occasion to occupy the White House on the fourth of March next, it shall be surrendered to you in perfect order from garret to cellar."

Mrs. James K. Polk

At one of her receptions, a gentleman said to Mrs. Polk in a loud voice: "Madame, there is a woe pronounced against you in the Bible." Mrs. Polk, who like her husband was religious and church-going, inquired what he meant. "Well, the Bible says 'Woe unto you when all men shall speak well of you'," he replied.

Mrs. Polk, who was her husband's private secretary, was always beautifully dressed and set the pattern for other ladies who attended receptions at the White House. Describing the beautiful women at one reception, an eyewitness said their "milliners sent them forth in fit trim to challenge the rainbow for the exquisiteness and variety of colors in which they were decked, while on their heads and bosoms glittering brilliants recline like nestling glow-worms, darting forth rays of light in dazzling emulation."

At their receptions, President Polk stood in the center of the Blue Room shaking hands with everyone presented, and saying a few words to each. Behind him stood Mrs. Polk and near her an affable elderly lady in a white or black turban. It was the irrepressible Dolley Madison, then in her eighties, but still a "presence" in the White House.

Guards inside kept the long line of guests moving by saying, "Gentlemen who have been presented, will please walk forward to the East Room. Don't stop up the passage!"

After the Mexican War developed, interest in White House receptions increased. Sometimes Polk would give the latest news from the battlefronts. Other times Mrs. Polk would hush all conversation to allow some young wounded officer to give a first-hand report of the war.

To add poignancy to the end of hostilities, Mexico's assent to the treaty (which added 525,000 miles of territory including the states of California, Utah, most of Arizona, New Mexico and parts of Colorado and Wyoming to the United States,) arrived at the White House on July 4, 1848, the same day the cornerstone of the Washington Monument was laid.

Gas illuminations were installed in the White House during the last winter the Polks were there, but Sarah Polk insisted that the elegant chandelier using candles be retained in the reception hall. It was fortunate that she did, for on the first reception night that the gas lights were in use, they all mysteriously expired and the house was plunged into total darkness except for the heartening glow of the candles in the reception room.

Charlotte Russe was a favorite dessert on the Polks' dinner menus.

CHARLOTTE RUSSE

Whip one quart of rich cream to a stiff froth. To one pint of milk add six eggs beaten lightly; add sugar to make very sweet; flavor strongly with vanilla. Cook over hot water until it forms a thick custard. Soak one ounce of gelatin in a little warm water. When the custard is very cold, add the gelatin and the whipped cream, beating lightly. Line the bottom of a mold with buttered paper, line the side with sponge cake or ladyfingers fastened together with the white of an egg. Fill with the cream. Put in a cold place or on ice. To turn out, dip mold for a moment in hot water.

Popular though she was, it was not with regret that Sarah Polk retired from the White House with her exhausted and overworked husband. They went to their new home in Nashville, named "Polk's Ease" where the President died three months later.

When Zachary Taylor, the hero of the Mexican War, became president, his wife, Margaret Mackall Smith, was 61 and worn out from moving from military post to military post with him. So she preferred to live practically as a recluse at the White House. Hostess duties were fulfilled by their youngest daughter, blond, blue-eyed, 22-year-old Betty Bliss, wife of Major W. W. S. Bliss.

Zachary Taylor

Mrs. Bliss entertained often and well at state dinners, weekly morning receptions and delightful little tea parties. Once a week the Marine Band played on the South Lawn and President Taylor strolled over the grounds among the company, shaking hands and introducing himself to those along the way.

He had a folksy manner that made him approachable, and one contemporary writer said he "received you as a grandfather receives his children, and you left his presence forgetting that you had seen the President but only feeling that you had been talking to one of the kindest old souls in existence."

Foods served by Mrs. Zachary Taylor often reflected her Tidewater Maryland childhood, although she had traveled with her Army husband from post to post on the borders of the United States. This recipe is for Tidewater Maryland Crab Cakes:

CRAB CAKES

1 pound crab meat
1 cup bread crumbs
1 egg, beaten
1 or 2 dashes hot sauce
1 tablespoon minced onion
2 tablespoons mayonnaise
1 teaspoon salt
1/8 teaspoon pepper
1/2 teaspoon dry mustard
1/8 teaspoon paprika
1 teaspoon lemon juice and pulp

Mix in the order given. Make into cakes and fry to a golden brown in hot fat.

During Taylor's administration the East Room was newly carpeted and redecorated. At his first reception in this brilliant new setting, he was host to six Osage Indians, who like the Indians who made such a hit at Jefferson's parties, added color to the event.

At his New Year's Day reception, 1850, the atmosphere was so congenial and friendly that after members of the Diplomatic Corps had been received by the President, they ignored tradition by grouping themselves around him and shaking hands with the people as they filed by.

Millard Fillmore

The most brilliant event of the Taylor years in the White House was the spectacular reception on March 4, 1850, to celebrate the first anniversary of his Presidency. Because of the effort expended some grumbled that it was an obvious bid for a second term. However, the criticism was soon regretted because the following July, Taylor died, after participating in Independence Day ceremonies at the Washington Monument. There he suffered a sunstroke and was quickly dispatched to a concerned White House. Five days later he died after serving only sixteen months in office.

Vice President Millard Fillmore, described as the handsomest of all the Presidents, was sworn in to succeed Taylor. Mrs. Fillmore, who was in delicate health, had remained in Buffalo while he was Vice President, and did not join him at the White House until October. The daughter of a Baptist minister, she had been Fillmore's teacher before their marriage, and during the years he was in Congress, she spent much of her time studying French and music.

Their's was a "dry" administration for Fillmore was anti-alcohol, anti-tobacco, and anti-gambling. But they entertained often and well. They held a reception every Tuesday morning and a levee every Friday night while Congress was in session. They held a large dinner for thirty-six every Thursday evening, and a smaller dinner on Saturdays.

Mrs. Fillmore, who had suffered an ankle injury in a fall years earlier, often stayed in bed during the day in order to have strength for the evening's social function. She was assisted in her hostess duties by their 18-year-old daughter, Mary Abigail, named for her great aunt, Abigail Adams.

What Mrs. Fillmore liked best in the way of entertaining was to invite choice friends to a private dinner and spend an evening in the library after dinner talking and listening to music. Often Mary Abigail played the piano, and mother and daughter sang duets on these private occasions.

The State Dining Room in President Pierce's Administration.

Mrs. Fillmore established the first library in the White House, having gotten a $5,000 appropriation from Congress for the project.

Just before leaving the White House the Fillmores entertained his successor, Franklin Pierce, at a dinner at which the guests included William Thackery and Washington Irving. Mrs. Fillmore caught cold at Pierce's inauguration and died in less than a month.

The administration of Franklin Pierce began on a note of gloom. Two months before his inauguration, the Pierces' third and last son died in a train accident. Grieving Mrs. Pierce, 47 and frail, with delicate features, ivory skin, black hair and large dark eyes, could not face up immediately to the duties of First Lady. She sent as her substitute Mrs. Abby Kent Means, a long time friend and second wife of her uncle.

It was a sparse social program during that period. Finally, a year and ten months after his inauguration, Mrs. Pierce joined her husband at the White House and forced herself to attend the social functions scheduled.

Franklin Pierce

She presided at weekly state dinners for thirty-six guests, morning and evening receptions, and the traditional New Year's Day receptions at which she wore deep mourning as she received with the President.

The state dinners were stiff and formal affairs, but they were of much interest to the socially-hungry capital. The ladies came dressed in their fanciest hoop-skirted gowns with their hair in corkscrew curls. The gentlemen wore sideburns and beards. At each plate at the dinners was a saucer-size bouquet of wired japonicas from the Government greenhouses, framed by a lace paper doily. The dinners were generally without wine, for Pierce, who had had a problem as an alcoholic earlier, had become a leader of the temperance movement.

Their evening receptions were the equivalent of the drawing rooms of foreign countries. After being received by the President and First Lady, the guests passed through the Green Room to the East Room where they mingled and moved like the ever-changing figures in a kaleidoscope.

No refreshments were served and many of the gentlemen, after escorting the ladies home, gathered at restaurants to comment on the evening over champagne, canvasback duck and venison steaks.

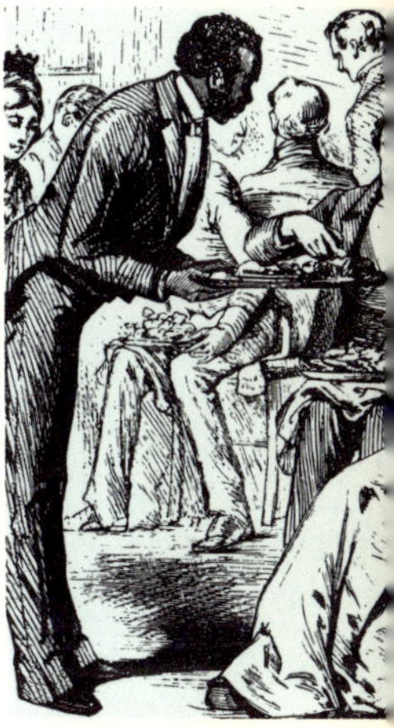

Tea party in President Taylor's time.

Pierce's last levee was attended by many more than usual, and the crush of guests was so great, one Colonel Lee who was a member of the city council and clerk in the Pension Office, seized a man near him by the collar and charged him with picking his pocket.

The gentleman so accused was a highly-respected merchant from Alexandria, Virginia, who went the next day to the Pension Office to convince Lee that he really had not attempted to steal his pocketbook. Lee refused to be convinced, and when the merchant struck him with his cane, Lee whipped out a pistol and shot him dead.

Dancing and gayety were restored to the White House by President James Buchanan of Pennsylvania, a bachelor. His official hostess was his 24-year-old niece, Harriet Lane, a willowy, blue-eyed blond who

An assembly in President Buchanan's Administration.

had been trained much like a princess when she lived with her uncle, then Ambassador to the Court of St. James.

She filled the once gloomy mansion with bright flowers—had an elegant conservatory built—and spent $20,000 buying new American-made furniture for the mansion.

It was a period when the rift between the North and the South was widening, however, and Harriet walked a tightrope of diplomacy to remain neutral and friendly with both sides, entertaining both at the same time.

James Buchanan

President Buchanan sent out five-by-seven inch card invitations to state dinners. However, no cards were issued for evening receptions and great crowds came. Miss Lane sent out invitations in her own name to daytime receptions and teas.

Some of President Buchanan's favorite recipes, served at the White House are:

PENNSYLVANIA DUTCH STUFFED SHOULDER OF PORK

5 to 6 pound shoulder of pork boned
salt
4 cups fresh sauerkraut
flour
1 tablespoon brown sugar
Garnish of cinnamon apples, tiny boiled carrots, mashed potatoes

Sprinkle the meat inside and out with salt and pepper. Fill the shoulder with the sauerkraut which has been drained. Sew or skewer the opening. Score the top lightly in a diagonal pattern as for baked ham. Dredge with flour and brown sugar.

Preheat oven to 450 degrees F., and start the roast in this for 10 minutes. Reduce heat to 350 degrees and bake 30 to 35 minutes to the pound, basting occasionally. Half an hour before the roast is done, pour off most of the drippings. Then place remainder of the sauerkraut in the roaster around the meat and complete roasting. Garnish the platter with cinnamon apples and mashed potatoes; add boiled carrots.

CINNAMON APPLES

8 Rome Beauty apples (of equal size)
1 cup sugar
2 cups water
½ cup cinnamon drops

Pare and core the apples. Boil sugar, water and cinnamon drops. When cinnamon drops dissolve, slowly add the apples, one at a time. Cook over low flame until tender, then remove apples from syrup. Boil syrup until it falls heavily from a spoon, then pour over the apples.

Harriet Lane Johnston

GOOSEBERRY TARTS

1½ cups flour
1 teaspoon salt
½ cup shortening
3 tablespoons water

Make pastry in usual way. Roll out on floured board. Fit pastry over the backs of fluted tart pans. Lay tart pans upside down on a baking sheet. Bake at 475 degrees F for eight to ten minutes.

FILLING

2 cups gooseberries
1 cup boiling water
⅔ to 1 cup sugar (according to the acidity of fruit)
2 tablespoons cornstarch
1 tablespoon butter

Cook gooseberries and water until tender. Drain off the juice and add sugar and cornstarch, then add fruit and cook ten minutes. Stir in butter, chill and pour into baked tart shells.

The two most spectacular social events in the Buchanan administration occurred in the same year. One was the visit of the Prince of Wales, later King Edward VII, son of Queen Victoria. He was entertained lavishly, and following a state dinner in his honor the Marine Band played a new song, "Listen to the Mocking Bird," which was dedicated by its composer, Alice Hawthorne, to Miss Lane.

A reception was given for the Prince, and people from all walks of life—ladies, gentlemen, officers, workmen, children and their nurses came to see the royal visitor. As each person passed, the President shook his hand, the Prince bowed his head, and a few ladies coaxed him into shaking their hands. When the crush of the crowd threatened to become too great, the Prince left the reception room and went to an upstairs window to be cheered by the throng outside.

The second event of exceptional interest was the reception for the Japanese diplomatic staff, the first delegation from that country, which arrived in May. It took place in the East Room and the doors were opened at 12 o'clock noon. Three Japanese ministers entered, accompanied by the Secretary of State and interpreters.

Very gracefully the ministers walked toward the President, standing at the end of a double row of Army and Navy officers. The ministers, in their colorful silk robes, bowed three times to President Buchanan, and then asked to be presented to Miss Lane and the cabinet wives, the only American women considered by them of sufficient rank to merit meeting. To the ladies, they bowed three times also in a colorful, graceful ceremony.

They brought with them twenty-eight boxes of gifts, among them screens, kimonos, lacquered ware, saddles, beautifully embroidered and embossed in gold and silver, a tea set inlaid with golden pearls, and a handsome cabinet that stood for 42 years in the Green Room as a memento of that historic visit.

As Buchanan's term came to an end, the young government, still less than a hundred years old, was facing its gravest crisis. To meet the challenge, America produced a charismatic figure who would next inhabit its Executive Mansion, Abraham Lincoln.

Abraham Lincoln

chapter 10

Wartime Entertaining

The threat of civil war hung heavily over the White House when President Abraham Lincoln and his petite, plump wife, Mary Todd, took up residence there on March 4th, 1861. But this threat did not disturb Mrs. Lincoln, who had social aspirations on a par with Julia Tyler.

She entered Washington with great ambitions, and she had the temperament and background to carry them out. The daughter of the president of the Bank of Kentucky, Mrs. Lincoln had been well-educated in an exclusive finishing school where only French was spoken and where the social graces were practiced as well as taught.

She had enjoyed a three-year social whirl in Springfield, Illinois, where her sister was married to the son of the Governor. Mary saw something uncommon about the promising young lawyer, Abraham Lincoln, and consented to be his wife.

Whereas Mary Lincoln was determined to reign as the nation's foremost hostess, her husband had a genuine liking for people and wanted to extend to them the hospitality of his house. So, the two supported each other during one of the most spectacular and socially-active administrations ever to occupy the White House.

The first reception the Lincolns gave at the White House occurred four days after they moved in, and it was a "come one, come all" affair. It seemed as if everyone did come that raw night on March 8, 1861.

An hour before the doors of the house were opened, the driveway was blocked with carriages, and the sidewalks and approaches to the White House were thronged with people coming to pay their respects to President and Mrs. Lincoln.

When the door opened at eight o'clock, the surge forward was so intense it was almost impossible for those in first to get out. From eight until half-past ten o'clock, the President, dressed in black and wearing white kid gloves, shook hands with the admiring thousands without taking a moment to rest. Beads of perspiration rolled down his face. He gave the double handshake, clasping the men with the right hand and the ladies with the left to speed the movement of the multitude.

At his side stood Mrs. Lincoln, her dimpled face smiling. She was handsomely dressed in a magenta water-marked silk with a lace cape and pearl necklace. Her abundant dark hair was worn with a half wreath of red and white japonicas. The skirt of her dress was bouffant over a hoop and stood out so that it was almost impossible for anyone to get closer than three feet.

After the guests had been received by the President and First Lady in the Blue Room, they passed through the "lovers' retreat" as Lincoln called the Green Room, into the huge East Room to inspect the colorful new carpeting. Lincoln had described it as a billowing ocean of green from which emerged pale pink roses as if floating on the surface.

As the rooms filled to capacity, it became necessary to open the windows and let some of the early arrivals climb out, because it was impossible to move against the incoming tide. For more than an hour the guests emerged in this fashion. There was little reason to linger for no refreshments were served. However, some people stayed on until well past midnight just to watch the conglomerate mob.

Not since Andrew Jackson's day had the White House been invaded socially by such an assortment of people. Backwoodsmen in heavy boots and flannel hunting shirts mingled with diplomats and members of Washington's social set in their finest silks and diamonds.

At half past ten, the Marine Band struck up "Yankee Doodle Dandy" and Mrs. Lincoln, with her hand on the arm of a former Congressman from Illinois, proceeded to the East Room for the traditional promenade. Behind her came the President with Elizabeth Todd Grimsley, Mrs. Lincoln's cousin, at his side. With him also was one of the Lincolns' young sons.

During the promenade it was easy for the crowd to see the six foot four inch President but only those along the promenade path could see his petite wife. When the President encountered a Virginian taller than himself, he asked good-naturedly: "Say, friend, does your head know when your feet get cold?"

Many of the guests who left their hats, coats and caps in the vestibule regretted it when it came time to go. The garments got mixed up and misappropriated, and whether intentional or accidental, many went home in the wrong hands. Some gentlemen tied their heads with handkerchiefs against the raw March wind rather than wear the greasy, dirty caps left behind. And the following day the newspapers carried a host of items under lost and found. Typical was the one that read: "Lost last night at the President's levee, a black ribbed cloth overcoat with velvet collar and facings."

That night, as throughout the Lincoln administration, souvenir hunters were out in force. They snipped pieces of fabric from sofas, lounges and chair covers, removed brackets from windows and tassels from curtains. On one occasion, several women were apprehended in the act of souvenir-snatching, and one collapsed in an embarrassed faint on the spot.

The day following that first levee, as receptions then were called, Mrs. Lincoln gave a reception for the diplomatic corps, and was sorely disappointed that not all of them came.

Before the month was out, the Lincolns gave their first state dinner, setting the hour at seven o'clock as advised by the State Department. It was in honor of the Vice President and Mrs. Hannibal Hamlin and members of the cabinet and their wives. It was a disappointment to Mrs. Lincoln, for it was not as gay as she had planned. Several cabinet wives sent excuses and did not come. And the President was called away from the table three times for urgent conferences.

She was happier, however, with the dinner for the diplomatic corps on June 4, 1861. One newspaper, The New York Herald, called it "most recherche and elegant" and "superior to anything" the guests had seen before. There were thirty-five guests and the dinner passed pleasantly with complimentary toasts, and, (a new feature) the exchange of elaborate snuff boxes, not only among the diplomats but the ladies as well.

The Lincolns continued to entertain on a grand scale even though the nation was in the grip of bloody conflict. There was a brilliant dinner for Prince Napoleon Bonaparte, son of Jerome Bonaparte, who came with his bride, Princess Clotilde, daughter of the King of Italy, in August. Twenty-seven guests sat down to an exquisite meal prepared by the White House staff rather than by a caterer as was the usual custom.

The first Japanese envoys to the United States call on President Buchanan.

Here are some recipes for dishes served during the Lincoln years, the original recipes having been adjusted to modern products and preparations:

SCALLOPED OYSTERS

2 doz. oysters (drained, save liquid)
¼ cup melted butter
2 cups coarse cracker crumbs (unsalted)
¼ teaspoon coarse black pepper
⅓ cup cream mixed with oyster liquid
2 tablespoons sherry
1 tablespoon Worchestershire sauce

Mix butter and cracker crumbs. Grease shallow baking dish with butter and put ⅓ of mixture on bottom. Add layers of oysters. Mix cream, oyster liquid, sherry, Worchestershire sauce and pepper. Pour one-half of sauce over layer of oysters. Add another one-third of crumbs, placing remaining oysters on top, and remaining sauce. Sprinkle remaining one-third of crumbs on top. Bake at 425 degrees for 10 or 15 minutes until crumbs are lightly browned.

SHERRY JELLY

1 box orange-flavored gelatin
2 envelopes gelatin dissolved in 1 cup cold water
¼ cup fresh lemon juice
1½ cups boiling water
1 cup ice water
1 cup Sherry (Amontillado)

Soak gelatin in cold water until dissolved. Empty the orange-flavored gelatin into a bowl. Pour in boiling water and stir until dissolved, then add dissolved gelatin, lemon juice and ice water; cool and add Sherry. Oil one large fancy mold or individual fancy molds. Pour liquid in mold and chill at least six hours before turning out. Serve with sweetened whipped cream flavored lightly with almond. Serves 12.

Mrs. Lincoln held two weekly receptions, one on Tuesday evening and one Saturday afternoon. The latter was supplanted in warm weather by a concert by the Marine Band on the South Lawn. The Lincoln family sat on the South Balcony on these occasions to receive all who chose to greet them there.

On January 1, 1861, a crowd estimated at five thousand turned out to shake hands with President Lincoln at the traditional New Year's Day reception, and although a line of police stood at each side of the entrance, there was much pushing and scrambling first to get in, and then to get out again, after having shaken hands with the man of the hour.

Mrs. Lincoln determined to restore dignity and order to their next reception and to show Southern senators and other sympathizers who had boycotted White House receptions after the war began, that she would not be insulted.

Mrs. Lincoln.

Her next reception, actually a ball, was held on February 5, 1862, ten months after the war began. Eight hundred guests were invited. The White House rooms were elegantly prepared with flowers, and the conservatory was used for the overflow of guests from the state rooms. Maillards, one of the outstanding caterers of the day, came down from New York to prepare the sumptuous supper.

Mrs. Lincoln, who had assuaged her disappointments as a hostess by building up an extensive wardrobe, wore one of her most beautiful gowns on this occasion. It was white satin with black lace flounces festooned with white and black ribbons. Her dark hair was arranged with a tight coil falling down her back and adorned with a bunch of crepe myrtle on one side.

The fact that she had two children sick and feverish upstairs was not allowed to detract from the festivities. She and the President both slipped away from the party once to look at the children but she concealed her anxiety. Fifteen days later, she lost her eleven-year-old, Willie, when he developed pneumonia.

Guests at the party arrived at nine o'clock, presenting their cards at the door. They went upstairs to the family rooms to leave their coats, and down again through the Blue Room to the East Room where President and Mrs. Lincoln stood at the door to receive them. There were a number of military uniforms in evidence and the one that rated the most attention was General McClellan, who was introduced to General Fremont, in civilian dress, by President Lincoln. Few members of the diplomatic corps were absent, and large numbers of Senators and Representatives from northern and border states were there.

The Marine Band played operatic airs. Also, "The President's March" and "Mrs. Lincoln's Polka," were played, although there was no dancing.

At 11 o'clock President Lincoln gave his arm to Miss Browning, the daughter of Senator Browning of Illinois, and Mrs. Lincoln, with Senator Browning, led the way to the State Dining Room for supper. The doors were locked, and for a few minutes no one seemed to know who had the key. There was good-natured jesting and jostling. Someone called out from the rear, "I'm in favor of a forward movement." "An advance to the front is only retarded by the imbecility of commanders," shouted another guest quoting a speech recently made in Congress. The Generals in line laughed as heartily as anyone. Finally the doors were opened and the crowd burst through to behold the table, a work of gastronomic art.

In the center of the main table that extended almost the length of the room was the Monroe mirror plateau and around it fancy pieces of confectionary.

Among the decorations and ornaments, according to one contemporary account, were a representation of a U. S. steam frigate with forty guns, all sails set and the flag of the Union flying at the main; a representation of the Hermitage; a warrior's helmet supported by cupids; Chinese pagoda; double cornucopias resting upon a shell supported by mermaids and surmounted

A White House reception.

by a crystal star, a rustic pavillion; the goddess of liberty, elevated above a simple shrine within which was a life-like fountain of water; a fountain of four consecutive bowls, supported by water nymphs; an elegant composition of nougat Parisenne; a basket of flowers and fruits mounted upon a pedestal supported by swans; plus thirty ornaments of cakes and candy.

On a side table was a large fort made of cake and candy and named Fort Pickens.

The menu was as lavish as the decorations. It included:

Stewed and Scalloped Oysters
Boned Turkey stuffed with Truffles
Paté de Fois Gras
Aspic of Tongue
Ham Venison Pheasant
Terrapin Chicken Salad
Charlotte Russe
Jellies Bonbons Biscuit Glace
Grapes Other Fruits

A ten-gallon bowl of champagne punch surrounded by little sandwiches was placed in the Green Room for before-and-after supper refreshments.

The festivities continued until three o'clock when the guests finally departed, and that in effect, was the finale of the period of entertaining by the Lincolns.

Such festivities at the White House at a time when boys were dying on bloody battlefields throughout the country provoked a storm of criticism in both the North and the South. Soon after, when Willie died, Mrs. Lincoln's anger turned to grief. She refused to be consoled. She went into seclusion. Band concerts on the South Lawn were discontinued. She refused to allow any music in the mansion, because Willie had liked music, and for the next few months, the White House was devoid of public entertaining.

On January 1, 1863, President Lincoln signed the Proclamation of Emancipation after holding the traditional New Year's Day reception. The weather was perfect, and thousands gathered at the White House to greet the President.

For several hours he and Mrs. Lincoln shook hands in the Blue Room. In the hall, the Marine Band played.

The President's arm became weary and when he went upstairs to sign the Proclamation, he had to hold his own hand still for several minutes to steady it before signing his name.

After that, Mrs. Lincoln resumed her hostess role, but entertaining was on a more limited scale. During this time Mrs. Lincoln was having personal problems. Her temper, that had always been difficult to keep in check, almost got out of control. She became extremely jealous of her husband, and formed an intense enmity toward Kate Chase Sprague, daughter of Secretary of the Treasury Salmon P. Chase of Ohio, and wife of Senator William Sprague of Rhode Island.

Mrs. Lincoln refused to invite Mrs. Sprague and her husband to a state dinner for the cabinet, and when the President's Secretary, John G. Nicolay, discovered the exclusion, he ordered the names reinstated. Mrs. Lincoln then turned her wrath on him.

At one of Lincoln's receptions in that period, General Ulysses S. Grant came unannounced, and the demand to see him was so great he was forced to stand on a sofa. Mrs. Lincoln, recognizing she had a social lion in tow, wanted to arrange a state dinner in his honor, but he was nowhere to be found. He sent word that he had seen enough of show business.

After Lincoln's second inauguration in March 1865, he and Mrs. Lincoln opened the White House to the public, and for four hours stood and shook hands with fifteen thousand persons. The President's hand developed blisters and bruises, but he went relentlessly on.

The next month, April 1865, the East Room where the tall, awkward President had welcomed so many admirers, became a funeral setting. He had been assassinated while attending a performance at Ford's Theater.

Mrs. Lincoln, now without an understanding husband to steer her through her difficulties, became more unable to cope than ever. She remained in the White House for five weeks after her husband's death, and it was June before the new President moved in.

Much personal unhappiness marked the Lincoln era, consummated by the tragedy of his assassination. Little wonder that some historians have called the Great Emancipator a "man of sorrow." But a new day was dawning at the White House.

Andrew Johnson

chapter 11

Post-War Entertaining

It was a gloomy, ill-kept White House into which President Andrew Johnson of Tennessee, moved his family in June, 1865. The mansion was in a state of ill repair, for troops had been quartered in the East Room during the Civil War. But the Johnsons had to make do with what they found until Congress appropriated $30,000 for redecorating and restoring the mansion in December of that year.

Mrs. Johnson, who had taught her husband to read and write and helped him rise from a backwoods tailor through the state legislature, Governor of Tennessee, both houses of Congress, the Vice Presidency and now the Presidency, was 55 years old and an invalid. She stayed in her room on the second floor of the White House while their two daughters, Martha Patterson, wife of a Senator, and Mary Stover, a widow, shared duties as official hostess for their father.

Mrs. Patterson, who drew the major role as hostess, was familiar with the White House, for she had spent weekends there during the Polk administration while she was a student at the Academy of Visitation in Georgetown. The Johnsons, however, had no social ambitions, and Martha Patterson said, "We are a plain people from the mountains of East Tennessee, called here for a short time by a national calamity, but we know our position and shall maintain it."

Social entertaining at the White House under Mrs. Patterson's leadership once again resumed the brilliance of pre-war years, and it again became the popular place to go for the social elite as well as the common folk. There was a homely dignity, free from ostentation, to the social functions during Johnson's presidency.

In the cookbook used by Mrs. Patterson was this recipe:

WHITE HOUSE PECAN PIE

1 cup unbroken pecan meats
1 cup dark table syrup
2 tablespoons butter
2 eggs
1 cup sugar
1 teaspoon vanilla

Cream the butter and sugar, add the table syrup, the beaten eggs, the pecans and vanilla. Beat together well. Put in unbaked pie shell and bake in a slow oven (275°) for about thirty minutes. Serve with whipped cream.

President Johnson gave his first reception on January 1, 1866 before the White House had been renovated. Martha made things look bright and gay by covering the old carpets with linen and hiding the bare spots with fresh flowers. She let her three children roam its rooms, giving them an aspect of purity and cheer.

She spent an entire summer supervising the redecorating in an effort to stretch the $30,000 Congress appropriated as far as possible. At President Johnson's New Year's Day reception, 1867, all the state rooms were redone except the East Room and it was closed to the public. Guests in the Red, Blue and Green parlors did not see the rich new velvet carpets, however, for Martha had covered them to keep them from getting soiled. And to help facilitate the movement of the crowds in the three rooms, a bridge was constructed from a hall window to the North grounds for an exit.

Martha and her sister Mary were always decorously dressed, wearing high necklines at a time when the gowns of their distinguished guests were cut to show the neck and shoulders. At one White House party, a Frenchman carried a jeweled opera glass in his left hand and when a woman with bare, dazzling shoulders came within view he leveled his glass to an exact angle on the parts exposed.

The two sisters were at home to guests every Monday afternoon and the President held levees twice a month during the season Congress was in session. State dinners were held on Tuesday evenings at seven o'clock and the entertainment after dinner was an hour of conversation in the Blue Room.

An exception was the dinner given for Queen Emma of Hawaii, the widow of King Kamehameha IV. After dinner the White House was thrown open to the public so that all who wished might come and meet her.

The queen wore a low cut gown of black silk with a mauve ribbon at her throat, and ornaments consisting of a diamond brooch and a necklace of jet. A tiara of jet crowned her head, and attached to the crown was a veil of white lace.

One of Mrs. Johnson's rare public appearances in the White House was at the children's party the President gave to celebrate his 60th birthday, in the closing months of his turbulent administration.

It was on December 30th, and almost four hundred children whose fathers were in public life and were not bitter enemies—were there. The invitation was issued in the President's name and he received the young guests as they arrived, with Mrs. Patterson and his grandchildren standing beside him.

Mrs. Johnson came downstairs for the party and was seated in one of the satin and ebony court chairs. When the young guests were presented to her, she told them "My dears, I am an invalid."

There was dancing in the East Room—square dances, a few waltzes and polkas. All of Marini's dancing academy were invited and they presented fancy dances—the Highland Fling, Sailor's Hornpipe and Spanish Dances. After that, all the children lined up for the Virginia Reel and then marched to the State Dining Room for the real party—the refreshments of ices, cakes, candies and glace fruits.

This was served at President Johnson's birthday party:

PINEAPPLE GLACE

Carefully peel and remove all discolored portions of a fresh pineapple; cut into slices about half an inch thick. Place in a preserving pan a pound of crushed loaf sugar and a pint of warm water and boil for ten minutes; put in the sliced pineapple and boil 15 minutes longer. Remove the slices of pineapple and place on a sieve to dry either in the sun or near heat for half an hour. Boil syrup until reduced and quite hard, dip in slices of pineapple again and place them on a butter plate. Dust with caster-sugar. When dried and hardened put some fresh green leaves in a glass dish, add the slices of pineapple and serve.

Mrs. Johnson came downstairs on another occasion also involving children. It was Easter Monday and the Presidential grandchildren had spent the day before Easter in the kitchen dyeing eggs for their Easter egg roll on the South grounds. After the children and their guests had finished rolling their eggs, the East Room was opened to them and they raced up and down the room singing, shouting and playing games.

There was no let-up in entertaining in the White House while President Johnson underwent his impeachment trial in Congress. He was charged with violating the Tenure of Office Act by dismissing his Secretary of War Edwin M. Stanton.

The Congress had passed an act decreeing that the President had no right to dismiss any member of his cabinet who had been confirmed by the Senate. During the trial on Capitol Hill, Johnson held a diplomatic reception at the White House with his daughter, Martha, smiling at his side as they received the foreign ministers and their court. The verdict was thirty-five for impeachment, nineteen against, one less than the two-thirds majority needed for impeachment.

Although Johnson may have been unpopular officially, his personal popularity was attested by the large crowd that attended his last levee. The East Room was dazzlingly illuminated with mirrors above the mantels, reflecting the light from the crystal chandeliers. The doors to the Blue Room opened on the President as the Marine Band played "Hail to the Chief," and the President, who stood in front of a magnificent arrangement of flowers, abandoned the system of announcing names as he greeted the guests.

At his side was Mrs. Patterson in black Lyons velvet with white lace collar, and a long blond curl hanging down her back, the stylish coiffure of the day.

At this reception, General Butler, leading instigator of the impeachment proceedings, appeared and shook hands with the President and they exchanged the customary compliments. Both looked a little embarrassed. Johnson was obviously touched, and after the interchange, Butler disappeared into the crowd, as suddenly as he had appeared.

Following Andrew Johnson came another military man to the Presidency; a man blessed with a wife so alive and vivacious that entertaining in the illustrious household involved no more than allowing her natural enthusiasm to come bubbling to the surface.

Martha Johnson Patterson

Ulysses S. Grant

chapter 12

Twenty-Nine Course Dinners

A new era of elegance began at the White House with the arrival of General Ulysses S. Grant and his Missouri-born wife, Julia Dent Grant. Julia was fair and forty-three when she became First Lady and she had a flair for entertaining that set new precedents in the historic old mansion.

"My life at the White House," she once said, "was like a bright and beautiful dream and we were immeasurably happy. It was quite the happiest period of my life. Life at the White House was a garden spot of orchids, it was a constant feast of cleverness and wit, a co-mingling with men who were the brainiest their states and countries could send to represent them and women unrivaled anywhere for beauty, talent and tact. When Congress and society get in session, Washington is a mecca for brains and beauty."

Her years as an Army wife made Julia Grant adaptable and she readily adjusted to life in the White House. The fact that she enjoyed it had an impact on those who shared the hospitality she and the President offered.

The Grants gave a dinner at the White House every Wednesday night. There were usually thirty-six guests—the maximum the dining room would seat then, and sometimes there were twenty-nine courses with a break in the middle at which Roman punch was served. The guests could get up and walk around if they wished, before starting the second half of the gastronomic feast.

Mrs. Grant had an Italian steward, Melah, who was responsible for the lavish banquets that cost from $700 to $1500 each. But Grant was a popular President and could afford it. Congress doubled his salary, from $25,000 to $50,000 a year, and also appropriated him the money for a lavish redecorating of the White House.

A favorite centerpiece for the table at the Grant dinners was a silver ship which was made by the Mohawk Indians for the Centennial Exposition in Philadelphia and afterwards was presented to President Grant. The table was ornately decorated with flowers, and a

State dinner given by President Grant in honor of the Joint High Commission.

bouquet of flowers with streamers of satin ribbon lay at each lady's cover. A boutonniere lay at each man's plate.

The Grants had many important persons as guests of honor at their dinners. Once it was Prince Arthur, third son of Queen Victoria, who was given a dinner said to have cost $2,000, an astronomical sum in those days for only thirty-six guests. Another time it was King Kalakaua of the Sandwich Islands whose entourage stood behind his chair and inspected every morsel he put to his lips.

The most historic dinner, however, was the one on March 3, 1877, for President-elect and Mrs. Rutherford B. Hayes. It was a lavish affair with covers laid for thirty-six guests and six wine glasses at every plate. The State Dining Room was elaborately decorated with palms, ferns and blooming flowers, and an azalea ten feet tall stood behind the chair of Mrs. Hayes. The menu included

Consommè Imperiale
Bisque de Grevisse
Sherry

Woodcock Patties — Salmon

White Wine

Filet of Beef — Crawfish Pudding
Breast of Pheasant — Goose Livers

Roman Punch

Artichokes — Turkey

Champagne

Canvasback Duck — Warm Sweet Dish

Red Wine

Guests included every member of Grant's cabinet and the Chief Justice. At midnight, Grant whisked Hayes and the Chief Justice into the Red Room where, unknown to the other guests, Hayes took the oath of office with the Secretary of State, Hamilton Fish, as witness.

Grant did this because Inauguration day, March 4, fell on Sunday. At the official ceremony on Monday, March 5, Hayes took the oath again in public ceremonies. Hayes had won the presidency by one electoral vote over Samuel J. Tilden, the Democrat who received half-a-million more popular votes than Hayes.

Writing of the state dinners given by President and Mrs. Grant, one social chronicle said: "Conversation at a state dinner cannot be general. A Congressman filled up the official time between each mouthful by telling his next lady the exact things which his palate craved. He didn't like "French dishes" but he was 'fond of pork and beans as well as ice cream and canned peaches.' "

Mrs. Ulysses S. Grant

The elegant dinners given by the Grants were described by Emily Edson Briggs, a newspaper correspondent who left a record of social Washington during the reconstruction period in her book, "The Olivia Letters," as follows:

"In the beginning of the feast, fruit, flowers and sweetmeats grace the table, whilst bread and butter give only a Spartan simplicity to the 'first course' which is composed of a French vegetable soup, and according to the description by those who have tasted it, no soup, foreign or domestic, has ever been known to equal it.

"The ambrosial soup is followed by a French croquet of meat. Four admirably-trained servants remove the plates between each course, and their motions are as perfect as clockwork. These servants are clad in garments of faultless cut, which serve to heighten to the last degree their sable complexion. White kid gloves add the finishing touch to this part of the entertainment. The third 'course' of the dinner is composed of a fillet of beef, flanked on each side by potatoes the size of a walnut, with plenty of mushrooms to keep them company. The next course is dainty in the extreme. It is made up entirely of luscious leg of partridges, and baptized by a French name entirely beyond my comprehension. It will readily be seen that a full description of the twenty-nine courses would be altogether too much for the healthy columns of a newspaper to bear, so we pass to the dessert, not omitting to say that the meridian or noon of the feast is marked by the guests being served bountifully with frozen punch. As a general rule, wine is served about every third course. Six wine glasses of different sizes and a small bouquet of flowers are placed before each guest at the beginning.

"The dessert is inaugurated by the destruction of a rice pudding, it is such a pudding as would make our grandmothers clap their hands with joy. After the rice pudding, canned peaches, pears and quinces are served.

Then follow confectionary, nuts, ice-cream, coffee and chocolate, and with these warm, soothing drinks, the Presidential entertainment comes to an end, and the host and his guests repair to the Red Room, and after fifteen minutes spent in conversation the actors in a state dinner rapidly disappear."

Following is a recipe for the Roman punch served at the halfway mark of the Grant dinners, and another recipe for dishes they served:

ROMAN PUNCH

Add one tablespoonful lemon extract to two quarts of lemonade. Mix well and freeze. When ready to serve, pour one-half pint of brandy and one-half pint of Jamaica rum over each quart of frozen lemonade. Mix well and serve in tall glasses.

VEAL OLIVES

Slice as large pieces as you can get from a leg of veal; make stuffing of grated bread, butter, a little onion, minced, salt, pepper and spread over the slices. Beat an egg and put over the stuffing; roll each slice up tightly and tie with a thread; stick a few cloves in them, grate bread thickly over them after they are put in the skillet, with butter and onions chopped fine; let cook until done; remove and lay them on a dish. Make your gravy and pour over them. Take the threads off and garnish with eggs, boiled hard, and serve. Cut in slices.

The Grants held two receptions each week. One in the daytime and the other in the evening. The first was held on Tuesdays from two to five o'clock and was called Mrs. Grant's reception.

She invited cabinet and Senate wives to luncheon preceding the reception and they joined her in the Blue Room in receiving the guests. That was also the day the cabinet met, and the President, sometimes accompanied by members of the cabinet, would drop in at some point in the afternoon.

At the evening receptions, the President and Mrs. Grant received together and as a special honor to their closest friends, let them sneak in a side door and stand behind the receiving line and watch the masses pass in single file.

One of the most crowded of the Grants' receptions was held on New Year's Day, 1873. It was also an occasion when some of the ladies wore their costliest costumes. And in this era of elegance and show, it was not unusual for a fashionable woman's gown and accessories to cost $1,000.

On this occasion Mrs. Grant wore a dress of pearl gray silk with flounces trimmed with point lace and silk of a darker hue.

Describing the crowd that came that day, one historian of the period wrote:

"The Army, the Navy, the Diplomatic Corps and the Judiciary were out in full force. There were nice people, questionable people, and people who were not nice at all in the crowd. Every state, every age, every social class, both sexes and all human colors were represented. There were wealthy bankers, and a poor, blind beggar led by a boy; men in broadcloth and men in homespun; men with beards and men without beards; members of the press and of the lobby; contractors and claim agents; office-holders and office-seekers; there were ladies from Paris in elegant attire, and ladies from the interior in calico; ladies whose cheeks were tinged with rouge, and others whose faces were weather-bronzed by out-door work; ladies as lovely as Eve, and others as naughty as Mary Magdalene; ladies in diamonds, and

White stock, yellow and white single chrysanthemums, white carnations, yellow and bronze daisies arranged in a large Chinese Export bowl for a state room in the White House.

others in dollar jewelry; chambermaids elbowed countesses, and all enjoyed themselves."

Mrs. Grant was often assisted in entertaining by her teen-age daughter, Nellie, who was the apple of her father's eye and the pet, as the only daughter in a family of three sons.

President Grant loved these:

SPICED OYSTERS

Bring to a boil fifty large oysters in their liquor salted lightly. Remove from fire and skim off the top, take out the oysters with a skimmer and place in a bowl to cool. To the liquor add one-half pint of vinegar, one small red pepper broken into pieces (do not use the seed), one whole cinnamon, one-half teaspoon each of mace and cloves and pepper to taste. Boil five minutes and pour while warm over the cooled oysters. Serve cold with thin slices of lemon.

Nellie Grant

The most brilliant social event in the White House during the Grant administration was the wedding of 19-year-old Nellie to Algernon Charles Frederick Sartoris, the handsome 23-year-old Englishman she had met on shipboard returning home from a visit to Europe.

The wedding was held in the East Room on May 21, 1874, and the President and Mrs. Grant sent invitations to two hundred persons including the highest-ranking officials in government and close family friends, many of whom were famous as warriors, statesmen and millionaire industrialists. Neither the name of the bride nor bridegroom was mentioned in the invitation which read:

"The President and Mrs. Grant request the pleasure of your company at the marriage of their daughter, at the Executive Mansion on Thursday, May 21st, at eleven o'clock a. m."

However, the names were not necessary for everyone knew that bouncy, brown-eyed Nellie and Algernon had waited for eighteen months for her father's consent to the marriage.

The wedding day dawned, a glorious spring morning with the fragrance of magnolia and catalpa heavy on the air outside the White House. Inside the fragrance of thousands of blossoms of tuberoses, spirea, lilies-of-the-valley and other flowers filled the East Room, which had been decorated by the Grants earlier. It had been given the full neo-Greek treatment with tall pillars jutting from the frescoed walls and ornate false beams splitting the ceiling into three sections. Each section was dominated by a colossal $1,800 chandelier made up of thousands of pieces of cut glass.

A dais at one end of the room was surrounded by plants and evergreens and over it a floral arch supporting a large bell composed of snow-white blossoms. Hanging from the ceiling just behind the bell was a horizonal band of flowers, and near each end of that was a small floral circle, one bearing the bride's initials and the other the bridegroom's. White satin ribbons formed an aisle to the altar and guests were grouped behind the ribbons.

As the Marine Band played the wedding march, the bridal party entered. The bridegroom set a precedent by carrying a bouquet of orange blossoms and tuberoses with a center of pink buds from which a tiny flagstaff protruded. There was one word on its banner: "Love." The best man was Nellie's handsome older brother, Colonel Fred Grant.

She had eight bridesmaids, all dressed in white corded silk with overdresses of white illusion and wide white sashes. Four carried blue bouquets and four carried pink. The bride wore a $5,000 white satin dress with wavy horizontal lines of delicate round point lace that ran across the front of the skirt and was intertwined with white flowers, green leaves and miniature oranges about the size of acorns.

Her bridal wreath was of white flowers and green leaves from which flowed a white tulle gossamer veil that enveloped the bride and extended to the floor. Her bouquet was choice white flowers, and she carried a pearl fan with lace cover.

A breakfast was served in the State Dining Room after the ceremony and this was as elaborate as anything the Grants ever gave. In the center of the table was the white bride's cake crowned with delicate flowers and a floral ribbon extending to each end of the table. On these ribbons and on trays of flowers at each end of the table were tiny American flags labeled "Success to the President," "Success to the Supreme Court," "Success to the Army," and the like. Around these were pyramids of nougat candy and outlining the edge of the table was a delicate line of flowers.

The menu at each place was printed on rich satin and tied with white ribbon. It read:

Soft Crabs on Toast
Gateaus garnis de Crabes et Champignons
Sauce a la Creme
Croquettes of Chicken with Green Peas
Cotelettes d'Agneaut Sauce a la Tartare
Aspic de langues de Boeuf a la Moderne
Woodcocks and Snipes on Toast
Broiled Spring Chicken
Salade Sauce Maillenaise Bride's Cake
Strawberries with Cream
Charlotte Russe Broques en Bouche
Corbeils glaces a la Jardiniere
Gateaux de Trois Freres
Epigraphe la fleur de Nelly Grant
Pudding a la Nesselrode Sauce a Gelee
Blamangea la Napoleon
Plombieres garnies de Fruits a Fleures Glaces
Ice Cream Water Ices
Small Cakes Punch a la Romaine
Coffee Chocolate

This was on the menu at Nellie Grant's wedding breakfast:

BROILED SNIPES ON TOAST

Pick, singe and draw eight fine snipes, remove the skin from the heads; split birds in halves without detaching the parts and place on a dish. Season with a pinch of salt, one-half pinch of pepper and one tablespoonful of oil. Put them to broil with the bills stuck into the breasts, and allow to cook four minutes on each side. Put eight slices of hot toast on a hot dish and arrange the snipes on them, spread one gill of butter on top. Decorate the dish with watercress and serve.

As Nellie and her husband left the White House for a train to New York from which they sailed to a new life in England, the chimes of the Metropolitan Methodist Church, whose minister had performed the ceremony, rang out with Mendelssohn's Wedding March, "The Wedding Pearl," "Hail Columbia," "God Save the Queen," "Then You'll Remember Me," "Auld Lang Syne," the Grand March from "Tannhauser," and "Home Sweet Home."

The wedding had been a spectacular event, the equal of which was not seen in the old mansion for many years to come. It reflected the warmth and effervescence of the Grants, a couple whose social training in the Army stood them in good stead in the most demanding social milieu in the country.

Nellie Grant whose marriage in the White House was an outstanding social event during the Grant Administration.

Rutherford B. Hayes

chapter 13

Hospitality Warm and Dry

Sunday night in the private parlor of the White House during the Hayes administration included hymn singing. Here the Secretary of the Interior, Carl Schurz, plays the piano.

There was a remarkable change in the atmosphere of White House entertaining when Rutherford B. Hayes, of Ohio, succeeded General Ulysses S. Grant as President.

Dinners and receptions were as elaborate and lavish as before, but no wines, liquors or Roman punch were served. Wine glasses were banned from the table. This caused some sharp criticism from guests. Some promptly dubbed the First Lady, who got the blame for the ban on liquor, "Lemonade Lucy," but it did not diminish her popularity. Actually, it was President Hayes who enforced the ban, for he was as stringently opposed to strong drink as was his gentle, religious wife.

When the two Russian Grand Dukes, Alexis and Constantine, were honored at a state dinner by the Hayes on April 19, 1877, the Secretary of State pleaded that the liquor ban be lifted for this special event and that wines, at least, be served.

Since the Grand Dukes were accustomed to wine, he argued, it would be rude not to serve it. Some of the guests thought he had won a partial victory when they were served at the dinner. On the menu were orange cups filled with frozen punch with the taste of old Santa Croix rum. Later they boasted that Hayes had served spirits.

However, President Hayes had the last laugh. He wrote in his diary: "The joke of the Roman punch oranges was not on us but on the drinking people. My orders were to flavor them rather strongly with the same flavor that is found in Jamaica rum. This took. There was not a drop of spirits in them."

The entertaining during President Hayes' term was as extravagant as that of any of his predecessors. A single reception cost $3,000, and the series of state dinners were on a similar scale. Mrs. Hayes gave many luncheons for ladies and every night was reception night. She considered herself ready to receive all who wished to come, and these informal evenings, one chronicler reported, "were bright and cheerful and of good report; (they) became the distinctive social feature of Mrs. Hayes' regime."

Mrs. Hayes, the daughter of an Ohio doctor, was well-educated, and had been the only girl student at Ohio Wesleyan University. Later she was graduated from the Cincinnati Wesleyan Woman's College. Her husband was an orphan who worked for his education, earning his law degree at Harvard.

They spent their early-married life in Cincinnati, where three of their eight children died in infancy. When they moved to the White House, they continued the same close family communion and religious practices they had observed earlier. Mrs. Hayes brought morning worship to the White House, just after breakfast, and cabinet members and Congressmen joined in their Sunday evening hymn singing there.

At her first afternoon reception Mrs. Hayes invited the wives of all the cabinet members to receive with her. After that they assisted her singly until the final reception of the season when she was surrounded once again by all of them.

Mrs. Hayes loved flowers and often wore white camellias surrounded by rose geranium leaves in her dark hair. She took a great interest in the White House conservatories and had a rose house and a violet house constructed outside the State Dining Room. During one state dinner, she had the long windows between the dining room and the conservatories opened, so that guests at the table could view the long vista of blossoms.

Flowers were used profusely also. At the second state dinner given by the Hayes, the table, set with thirty-six covers, was decorated with bouquets of pink and red roses, and the oval Monroe mirror plateau was bordered with calla lilies separated by clusters of green.

The card at each place bore the national coat of arms embossed in gold, and the name of the guest inscribed in old English. There was a hand bouquet of flowers for each lady and a boutonniere half-hidden in each man's napkin. A decanter of water, rather than wine glasses, was beside each plate.

One of the outstanding social events at the White House during the Hayes Administration took place on December 30, 1877, when the President and First Lady celebrated their twenty-fifth wedding anniversary. It was the first silver wedding anniversary to be observed in the White House and the celebration lasted two days. A clerk wrote the invitations, and across them President Hayes himself inscribed "I hope you will come."

On that Sunday afternoon, he and Mrs. Hayes renewed the marriage vows they had taken a quarter of a century earlier. The same preacher who married them, Dr. L. D. McCabe, of Delaware, Ohio, conducted the ceremony again.

Mrs. Hayes wore the same white satin dress and slippers she had worn on her wedding day. The bridal couple were surrounded by their five children and the friends who had attended them when they were married. Among them were Mr. and Mrs. John W. Herron, of Ohio, whose daughter was later to become a First Lady.

Following the services, the Herrons' infant daughter, named Lucy Webb in honor of Mrs. Hayes, was christened, and then, the President presented his daughter, Fannie, and youngest son, Scott Russell, for baptism. Among the guests was the Herrons' seventeen-year-old daughter, Helen, who became mistress of the executive mansion when her husband, William Howard Taft became President.

Following the vow-renewal ceremony and the baptisms, the Hayes led their guests to the dining room where dinner was served.

The next night the public reception celebrating the silver anniversary was held. President and Mrs. Hayes descended the stairs to the strains of "Mendelssohn's Wedding March." They were followed by the house guests, and all proceeded to the East Room where they received in front of the East windows. Mrs. Hayes did not wear her wedding gown this time but a white silk with draperies of white brocade, the heart-shaped neck filled with tulle, and the half-sleeves ending with lace. Her only ornament was a silver comb fastening back her heavy bands of glossy black hair.

The guests included many of the people who had attended the wedding in 1852, among them Mrs. Hayes' former teachers and Mrs. Mitchell, the President's niece, who as a little girl was the bride's attendant and held her hand during the ceremony.

The President had notified the guests that they did not wish any anniversary gifts and the only one received came from officers of the Twenty-Third Ohio Volunteer Infantry which the President once commanded. It consisted of a silver plaque against a black velvet background enclosed in an ornamental ebony frame.

A token of appreciation for the kindness received from Mrs. Hayes, the plaque was inscribed to her as "The Mother of the Regiment" and continued:

"To Thee, 'Mother of ours,' from the Twenty-Third O.V.I. To Thee, our Mother, on thy silver troth, we bring this token of our love. The boys give greeting unto thee with burning hearts. Take the hoarded treasures of thy speech, kind words, gentle when a gentle word was worth the surgery of an hundred schools to heal sick thought and make our bruises whole. Take it, our Mother; 'tis but some small part of thy rare beauty we give back to thee, and while love speaks in silver, from our hearts we'll bribe Old Father Time to spare his gift."

Above the inscription was a sketch of the log cabin which served as Colonel Hayes' headquarters in the valley of Kanasha during the winter of 1863-64, and the tattered and torn battle flags of the regiment.

The White House was the setting for another romantic event during the Hayes administration. The President's twenty-seven-year-old niece, who had lived with them most of the time since her mother's death twenty-one years earlier, was married on June 19, 1878 to General Russell Hastings, who had served with Hayes in the Twenty-Third regiment during the Civil War.

He had met Emily during the war in Columbus, Ohio, but it was not until he called at the White House to visit the President that she caught his eye romantically. He had been a widower for three years then.

The wedding was held in the Blue Room, which was banked with geraniums, fuschias and foliage plants. In front of the windows hung a wedding bell of white roses with a rope of green festooned on each side and wreaths of roses and pansies hung beneath the wall lights.

The Marine Band struck up the wedding march at seven o'clock in the evening, and the bridal party entered the Blue Room. First the President and the bride's married sister, followed by the bridegroom and Mrs. Hayes. The bride entered on the arm of her father. She wore a gown of ivory brocade with a square neckline filled in with tulle and orange blossoms. A long train bordered with deep pleating of plain silk was caught at the top with a wreath of orange blossoms on her head.

An elaborate supper was served afterwards in the family dining room where a bride's cake decorated each end of the long table. Each place was marked by a bouquet of flowers, with baskets of fruits, pyramids of ices and flowers, and plates of bonbons in the center of the table. There was no champagne, but toasts to the bridal couple were drunk in tea, coffee and lemonade.

Mrs. Hayes' no-liquor policy won her as much praise as criticism. Much of the applause came from the Women's Christian Temperance Union, which presented a life-size portrait of Lucy Hayes by Daniel

Huntington to the White House for its collection.

During their second year in the White House, President and Mrs. Hayes began the practice of issuing invitations for the receptions for the diplomatic corps, the Army and Navy, the Judiciary, and Congress. About one thousand were invited to the first reception in February, 1878, and the White House was decorated from top to bottom for the event.

Guests were permitted not only in the state rooms but also in the upstairs room and the library. Even some of the offices were used for serving refreshments. These included sweetbreads, bouillon, patties, salads, creams, ices, cakes, coffee, sweet meats and others which were traditionally served at White House receptions except the forbidden wines and punches.

Every Thanksgiving Day President and Mrs. Hayes invited the White House secretaries, clerks and their families to a dinner which he called his office family reunion. Dinner was served at an early hour for the convenience of the children (who were given souvenirs), and the event was almost as ceremonious as a state affair.

Each year the President escorted a different lady to the dining room, and after dinner, all gathered in the Red Room where Mrs. Hayes played games with the children, joined by the younger Hayes children, Fannie and Scott. The games included "Kitty Wants a Corner" and "Pass the Button." After the younger ones grew tired, the adults sang hymns around the piano, played by Mrs. Hayes.

Each year she bought a Christmas gift for everyone who worked at the White House, and about noon on the holiday, the entire staff—office and household—gathered in the library where the wrapped gifts were piled in the middle of the floor. President and Mrs. Hayes lifted them one by one, read the names, and the gifts were delivered by Fannie and Scott, who scampered like a couple of Santa's helpers to place them in the right hands.

The Hayes' brought to the White House grounds that delightful old capital custom of a public egg rolling for children on Easter Monday. For years the children had rolled their eggs on the sloping Capitol grounds, but one Easter Monday they were chased away by police and messengers from the House and Senate who said they did not want their lawn disfigured by egg shells.

President Hayes, riding by at that moment, saw the children crying and stopped to see what was wrong. Then he invited the children to come and roll their eggs in his backyard. When the gates were opened at eleven o'clock, the children were waiting to go in. Ever since, it has been a tradition for Washington children to go to the White House on Easter Monday.

Egg rolling in the early years was a challenging competition. The children divided into groups of eight or ten and rolled their eggs down a slope. Those whose eggs reached the bottom intact, after eliminating the others by breaking them enroute, became the winners.

James A. Garfield

Several brilliant functions were held at the White House during the closing months of President Hayes' term. One was a dinner in honor of former President and Mrs. Grant, and another was for incoming President and Mrs. James A. Garfield.

President-elect Garfield's aged mother was a guest at the dinner in his honor but she was unable to participate for very long. So she retired early, and Fannie Hayes, by then a teen-ager, took her place at the table.

President Garfield, who succeeded Hayes in March 1881, hardly had time to make a social record in the White House before he died from an assassin's bullet six months after taking office. But the White House was a lively place during the four months he, his wife, their five children, and his mother lived there.

For his arrival on March 4, the White House was gaily decorated with lines of streamers and signal flags running from tree to tree across the semicircular drive to the entrance, and across the lawn itself. The columns on the North Portico were decorated with evergreens, and in the pediment of the portico was a large glass star in red, white and blue.

During the years Garfield was in Congress, Mrs. Garfield had entertained frequently and her receptions were better attended than those of any other Senate wife. As First Lady she applied her skills as a gracious hostess to making the White House receptions enjoyable events.

Six days after moving into the mansion she and the President gave a reception for the diplomatic corps. Garfield's comment about the evening was "it was very pleasant. C (meaning Lucretia, whom he called Crete) grows up to every new emergency with fine tact and faultless taste."

The next reception was for officers of the army and navy. The President's comment this time was, "The array was brilliant." On March 17, the Irish Societies celebrated St. Patrick's Day by marching around the White House and through the porte-cochere, with the President watching.

On April 2nd, the Garfields gave their first formal dinner. It was in honor of his old college president, Dr. Hopkins of Williams College. "A very pleasant party and a good dinner," the President commented later in his journal.

Mrs. James A. Garfield

Mrs. Garfield held a reception twice a week, and the President often assisted her, enjoying them as much as anyone.

The First Lady set as her goal refurnishing the mansion with authentic pieces, and spent her spare time in the Library of Congress looking into the history of the White House and its furnishings. This was an interest the President shared.

A true intellectual, Mrs. Garfield was the daughter of one of the founders of Hiram College. There she and Garfield were classmates, both preparing to be teachers. Later he taught Latin and Greek, eventually becoming president of the college.

At the White House the Garfields turned the family dinner table into a classroom for the educational development of their five children, Harry, eighteen, James, sixteen, Mollie, fourteen, Irwin, eleven, and Abram, nine. The two younger boys occasionally entertained their friends with bicycle races in the East Room—a new use for the room that had been the scene of many brilliant balls.

Mrs. Garfield had an attack of malaria in April and in the middle of June the family went to Elberon, New Jersey with the hope that the sea air might benefit her in her convalescence. She was making progress, and planned to accompany him the next month on a tour of New England.

On July 2nd, at the train station to begin his journey to a college reunion and then to meet Mrs. Garfield, the President was shot by an insane office-seeker. He was returned to the White House where he lingered in great pain for two long months before he died on September 19th.

When Vice President Chester Arthur, a man of discriminating taste, ascended to the presidency upon Garfield's death, he refused to move into the White House until it had been redecorated.

Chester A. Arthur

"I will not live in a house looking this way," he declared. "If Congress does not make an appropriation I will go ahead and have it done and pay for it out of my own pocket."

He had good reason for this attitude for a visitor to the mansion at that time described it as being "a place full of modern abominations in upholstery and garish gilding and all the rooms look staring, pretentious and Frenchy. The old port-wine colored mahogany sofas and chairs which were in the state parlors in Lincoln's time were better than anything that has come in their place. At least they were quiet and dignified."

Arthur cleared out the place, and five thousand persons gathered at the auction to bid for his cast-offs. He won his appropriation from Congress, and began with the East Room, which was redone with a ceiling in silver. The Blue Room was decorated in robin's egg blue, with ornaments in ivory on the hand-pressed paper. The Red Room walls were red with a ceiling of old gold, and a motif in the frieze of eagles and flags. An opalescent glass screen was erected in the front hall, also decorated with eagles and flags. And he installed the first elevator, and added two new bathrooms.

Arthur, who had been a widower for a year and a half before he became president, entertained elegantly and often. He moved the dinner hour back from seven o'clock to eight, and reduced the number of courses to fourteen, with eight varieties of wine.

Mrs. James G. Blaine, wife of the Secretary of State, said his dinners were "extremely elegant, hardly a trace of the old White House taint being perceptible anywhere, the flowers, the damask, the silver, the attendants, all showing the latest style in an abandon in expense and taste.''

He gave recognition to people of significant achievement in his day by inviting them to dinner at the White House, and often had fifty-four guests instead of the traditional thirty-six.

President Arthur's wife had been a talented singer and he loved music. His entertaining at the White House was marked by the performances of outstanding musical talent of his day. On Washington's Birthday in 1882, the Fisk Jubilee Troupe sang for his guests, and after a dinner in her honor, Christine Nilsson, the Swedish soprano, sang ''The Last Rose of Summer'' and ''Way Down Upon the Swanee River,'' accompanying herself on the piano in the East Room.

Arthur's sister, Mrs. John McElroy, widow of a minister, served as his White House hostess. She held weekly receptions from three to five o'clock in the afternoon, and asked up to fifty prominent women, including two former White House hostesses, Mrs. John Tyler and Harriet Lane Johnson, to assist her. Afterwards, the assistant hostesses would gather on the second floor in the upper hall for tea and wafers. Sometimes the President would join them for tea, and remain to chat.

At Arthur's first evening reception, former President and Mrs. Grant had just returned from a world tour and were the guests of honor. Mrs. Grant took a place in the receiving line.

One of Arthur's New Year's Day receptions was suspended after the Minister of Hawaii, then dean of the diplomatic corps, died suddenly in the midst of the gay throng lined up to greet the President.

Mrs. McElroy's last reception was so crowded that the Marine Band was swept from its position in the hall, and could not continue to play because of the great crowd. Some of the guests, including General Phil Sheridan, were helped by police to gain entrance through a portico window because the doors were so crowded.

The White House was the social style-setter in these years and the crowds thronged to bask in that leadership.

Grover Cleveland

chapter 14

A Young Bride Sets the Pace

Grover Cleveland, the first Democrat to occupy the White House in twenty years, was a forty-eight-year-old bachelor when he became President on March 4, 1885, to succeed Chester Arthur. His sister, Rose, a former teacher in an exclusive girls' school, came from the family home near Utica, New York, to serve as his hostess.

She served also as a chaperon during his White House courtship of his ward, Frances Folsom, the twenty-one-year-old beauty he married his second year in office. Miss Folsom and her mother were guests in the White House following Cleveland's inauguration, and they returned in June of the next year for the first wedding of a President in the White House.

Rose Cleveland planned and presided at her brother's first state dinner, in honor of the cabinet.

The decorations were a profusion of plants, flowers and art in wax.

At each end of the table were ornaments in white wax. At the east end, white swans with outspread wings upheld three fancy molds of jellied pate de foie gras. At the west end, the figures were eagles with the pate arranged like little square blocks attached by silver skewers to horseshoes above.

Interspersed the length of the table were glass and silver stands of conserves, bonbons and salted almonds. The service used was the ornately beautiful china of the Hayes administration decorated with the flora and fauna of the United States. Six wine glasses, a water tumbler and a champagne glass stood at each place, and salt cellars of cut glass with golden spoons and silver pepper stands were beside each plate.

President Cleveland gave each guest at state dinners a souvenir in the form of a ribbon. The ladies' ribbon was about three feet long and three inches wide, with a picture of the White House stamped in black on one end, the national colors, the coat of arms of the United States in gold at the other end, and the name of the guest and date of the dinner in gold letters between the two. The gentlemen received similar ribbons but of smaller size with only the coat of arms, name of guest and date of dinner stamped on them.

President Grover Cleveland, only President to be married in the White House, and his bride, the former Frances Folsom.

There was also a bouquet with a gilt bullet-headed pin at each lady's place, and a boutonniere of green with a single Bon Silene rose for each gentleman.

Rose Cleveland entertained at receptions and luncheons but at these she adhered to her temperance principles and no wines were served. Although she filled her White House role with charm and graciousness, she was pleased to be able to turn the role over to her brother's bride, and she could return to her career as a teacher and writer.

President Cleveland, who had kept his romance with his ward a secret even from her mother for a long time, gave the public the first inkling of what was to happen just four days before the wedding. He wrote the invitations that went to the family, cabinet members and the few friends who were to witness the event. Dated May 29, 1886, the invitations read:

My dear Mr. ....,

I am to be married on Wednesday evening at seven o'clock at the White House to Miss Folsom. It will be a very quiet affair and I will be extremely gratified at your attendance on that occasion.

Yours sincerely,
Grover Cleveland

He likewise supervised the decorations, and helped select the music played by the Marine Band.

The state rooms were profusely decked with flowers, palms and tropical grasses. The large windows at the South side of the oval Blue Room were hidden behind a foliage that touched the ceiling. Tiered below were various shaded plants accented by red geraniums and white lilies. The fireplace was filled with celias and small pink flowers and on the mantel was a bed of red roses with the monogram "C-F" in white. On the opposite wall was a bank of red roses and purple pansies with the date formed by yellow pansies. Above the door from the main corridor were the words "E Pluribus Unum" worked in blue immortelles on a deep red background.

The new Mrs. Cleveland is kissed by her mother following the ceremony.

At seven o'clock the Marine Band struck up Mendelssohn's "Wedding March" and the President came slowly down the staircase with his bride holding his arm. She wore a Paris gown of corded ivory satin trimmed with silk India muslin crossed in Grecian folds over the front of the high corsage and fastened in the folds of the satin at the left side. Orange blossoms and leaves edged the V of the bodice and extended down the left side of the draped skirt. The three-quarter length sleeves were also trimmed with orange blossoms.

The Reverend Byron Sunderland of the First Presbyterian Church performed the wedding ceremony, and the President's brother, the Reverend William Cleveland, pronounced the benediction. The bride's mother stepped forward and kissed her cheek.

As the Marine Band played the "Bridal Chorus" from "Lohengrin," the President led his beaming wife to the family dining room where the wedding supper was served.

It consisted of terrapin, breast of spring chicken, cold meats, salads, fish, pate de foie gras, ice cream molds, bonbons and fruits. The centerpiece on the table was a ship outlined in pinks, roses and pansies. It was named Hymen for the Greek god of marriage. The mirror of the Monroe plateau represented the lake on which it sailed. When champagne toasts were drunk, the young bride drank her toast in mineral water.

Each guest received a satin bag and box of bonbons as well as a piece of wedding cake in a little box on which had been painted the date and a spray of flowers.

Later that evening the President and his bride, who had been his ward since she was eleven, left under a shower of rice and a trail of slippers for a honeymoon at Deer Park, Maryland.

President Cleveland had the White House redecorated and painted for the new First Lady, and after they returned from the honeymoon she took up her duties with such enthusiasm and zest she soon became the nation's favorite pet.

Mrs. Cleveland began her social entertaining at the New Year's reception of 1887, and this was followed by a dinner for the cabinet on January 20th. She took special interest in the decorations for her parties, and this dinner showed her interest and good taste.

The table was decorated with a big ship of red and white camellias, the sails trimmed with green smilax. It sailed on a mirror lake bordered with camellias, rosebuds and tulips. On each side of the ship were stands of fruit and vases of long stemmed roses. At each end were two flat cushions of orchids, red and yellow roses, tulips and carnations.

At the state dinner for the Supreme Court in February, a great mound of Bon Silene roses edged with smilax was in the center of the table and upon it two open books fashioned from white immortelles with the words "Book of the Law" worked in with purple flowers. At each end of the table were two crossed swords of justice made of red and white carnations.

The Clevelands adopted a new style of invitations for official receptions, but it was unpopular. Senators, Representatives and the Judiciary were invited to each reception and received an engraved invitation at the beginning of each session listing on a single card the dates of the receptions. The invitation itself, on heavy white bristol board about six by four and one-half inches, read:

The President and Mrs. Cleveland
Request the pleasure of the company of
The Senators and Representatives in Congress
And the ladies of their families
On Thursday evening, January 13 and 27 and February 10
from 9 to 11 o'clock 1887
Jan. 13—To meet the Diplomatic Corps
Jan. 27—Reception to Congress and the Judiciary
Feb. 10—To meet the officers of the Army and Navy and Marine Corps
Feb. 22—The public reception

The diplomats were invited by a printed circular sent out by the Secretary of State. The Army and Navy were invited by the Secretary of War and the Secretary of the Navy, and the Judiciary by the Attorney General.

This method of inviting guests to the White House met with strong disapproval from those who had formerly received their invitations directly from the President, and subsequently, the Clevelands reverted to the old custom.

There was a new design for the place cards at the table. It consisted of an eagle with the shield of the United States on its breast, the claws grasping the arrows and olive branches, and the thirteen stars above its head.

In addition to the official receptions and dinners, Mrs. Cleveland had a full schedule of receptions, luncheons and teas of her own. Her invitations had the words "Executive Mansion" in silver at the top and read:

Mrs. Cleveland
requests the pleasure of the company
of Miss ....
at luncheon
Wednesday, January 12
at 1:30 o'clock 1887

At one of her luncheons the table was supplied with a dish of candy for every two guests, and beside it, saucers filled with salted almonds (for it was still before calorie-counting became popular). In the candy stands were sticks of chewing gum wrapped in fancy colored papers.

GRAPES GLAZED WITH SUGAR

Boil one cup of syrup as for candy. Add half a teaspoon of cream of tartar. Divide a few bunches of grapes into clusters, dip them into the candy and lay on slightly greased dishes to dry.

Four golden candelabra filled with tapers and capped with shades, and four silver candelabra lower than the others were spaced out on the long table. Alternating with them were bouquets of pink roses tied with heliotrope ribbon and bouquets of heliotrope tied with pink ribbon. Thirteen courses were served at the luncheon but no wine. In fact, Frances Cleveland was a "tee-totaler" and always had her wine glasses removed from the table at state dinners.

Frances Cleveland, who had graduated from Wells College the year before her marriage, held Saturday afternoon receptions so that shop girls employed the rest of the week could come and see the White House and shake her hand. They were the most popular social events she held, and the attendance grew until one afternoon the line waiting to be received stretched from the White House entrance to the Treasury Building in the next block. By count, nine thousand people passed through the Blue Room and shook her hand. Later she had to have both arms massaged as they were numb from so much activity.

In the fall of 1888, despite his wife's popularity, and much to everyone's surprise, President Cleveland was defeated in his bid for re-election. It was a blow to his young wife and she cried when she left the White House. But she told the staff to take care of everything in the house for "we are coming back just four years from today." And they did.

On their return they brought with them a baby girl, Ruth, who was just over a year old. In September, six months after moving back into the White House, a second daughter was born to the Clevelands.

Named Esther, she was the first child of a President to be born in the White House. In July 1895, a third daughter, Marion, arrived but she was born at Buzzard's Bay, Massachusetts, where the President and First Lady were spending the summer.

In May, 1893, the Clevelands gave a dinner for the Princess Eulalie of Spain, a member of the royal family. She was accompanied by her husband and nineteen Spanish nobles. Because of the great number invited, the dinner was held in the East Room. She arrived a little late and one of the ushers took her wraps at the entrance to save her the time of going to the library. Then to everyone's surprise she walked directly to the East Room and presented herself to the President who had walked toward her with outstretched hand. Some of the officials were awaiting her in the library on the second floor and planned to escort her down to the East Room for a ceremonious presentation, but her straightforward manner foiled their plan.

Cleveland's last Christmas in the White House was an especially happy holiday. A carnival was held in the library for his own three little girls and the children of cabinet members. A fir tree, decorated with tinsel, gilt and artificial snow, stood in the center of the room, its top touching the ceiling. The bright-eyed children cavorted around it, and after refreshments, little Ruth and Esther Cleveland who had been learning German, gave recitations. This was followed by the telling of Christmas stories and the singing of carols by the children.

Crowd gathers outside the White House for a glimpse through the windows of the Cleveland-Folsom wedding.

It was Benjamin Harrison, however, who had the first Christmas tree at the White House. He was President from 1889 to 1893, between Grover Cleveland's two terms in office. And it was on his first Christmas eve in the White House that he decorated the first Christmas tree in the mansion. It was placed in the oval room on the second floor and he was joined in the happy project by all the adult members of his family and his staff.

Benjamin Harrison

The following is the Harrison's Christmas dinner menu.

PRESIDENT HARRISON'S CHRISTMAS DINNER

Blue Point Oysters, half shell
Consommé Royal
Bouches à la Reine
Turkey   Cranberry Jelly
Potatoes Duchesse   Stewed Celery
Terrapin à la Maryland
Lettuce Salad   Plain Dressing
Mince Pie   American Plum Pudding
Tutti-frutti Ice Cream
Lady-fingers   Macaroons   Carlsbad Wafers
Apples, Florida Oranges, Bananas, Grapes, Pears
Black Coffee

FIG PUDDING

1 cup black molasses	2 eggs
1 cup chopped suet	1 teaspoon soda
1 cup milk	1 teaspoon cinnamon
3¼ cups flour	½ teaspoon nutmeg

1 pint figs

Mix together the molasses, suet, spice and figs, cut fine. Dissolve the soda with a tablespoonful of hot water and mix with the milk; add to the other ingredients. Beat the eggs until light, and stir in the mixture; add the flour and beat thoroughly. Butter two small or one large brownbread mold; turn the mixture into the mold and steam five hours. Serve with cream sauce.

The Harrisons were a big, happy family in the White House although his four years were interrupted frequently by sorrowful incidents. The family included the President and his artistic wife, fifty-seven-year-old Caroline Scott Harrison of Ohio, their daughter, Mary Harrison McKee, her husband and their two children, Mrs. Harrison's elderly father, a minister; her widowed niece, and Mary Scott Dimmick, a charming woman of about thirty who acted as social secretary to the First Lady. Frequent visitors to the mansion were the Harrisons' son, Russell, and his wife and their daughter, Marthena, so they were almost members of the household.

Mrs. Benjamin Harrison

The McKees' two children, Mary and Benjamin Harrison (called Baby McKee) were the delight not only of their grandparents but of the entire country. Their presence and activity were reported widely in the daily press, and the President, who was sometimes considered cold by outsiders, showed his true warmth when he played with his grandchildren.

Once he was outdoors with Baby McKee, who was riding in his goat-drawn carriage around the White House grounds when suddenly the goat ran through the gates and down the street with the dignified President, in frock coat and top hat, in hot pursuit.

The Harrisons entertained elegantly and Mrs. Harrison, who was accustomed to Washington social life from her six years as a Senator's wife, set something of a record in the use of plants and flowers in decorating the White House for official functions.

At one dinner alone she used five thousand decorative plants for the East Room and about a mile of smilax to festoon the chandeliers.

At the dinner on January 7, 1890, for Vice President Levi Morton and members of the cabinet, miniature groves of palms touching the ceiling were arranged in the corners of the East Room, and before the windows. The four mantels were banked with two thousand azaleas, eight hundred carnations, three hundred roses, three hundred tulips, nine hundred hyacinths, four hundred lilies of the valley, two hundred bouvardias, one hundred sprays of asparagus fern, forty heads of poinsettas and two hundred small ferns.

The centerpiece in the State Dining Room was a ship with the hull made from carnations, the deck of narcissus and the rigging of white acacia. The room was brightened by scarlet tulips and poinsettas on the mantel and in the window niches.

The menu was only a little less impressive. It included:

Green Turtle Soup
Raw Oysters   Boiled Salmon
Filet of Beef a la Jardiniere
Canvasback Duck
Ices in the shape of Roses and Chrysanthemums

A mirror view of the Red Room as it is today.

Mrs. Harrison's favorite flower was the orchid, and a wealth of them were used at the state dinner for the diplomatic corps on January 21, 1890. The oval centerpieces were wholly of orchids, and the bouquets for the ladies were pink orchids. Near the ends of the table which was shaped like a double L, to seat forty-eight guests, were floral hemispheres to signify the presence of representatives from all parts of the globe.

The most unique entertaining done by the Harrisons, however, was the birthday party the President gave on March 16, 1891 to celebrate "Baby McKee's" fourth birthday.

The guests were greeted in the East Room by the President and his grandson, who together led them to the dining room where fifteen high chairs encircled a round table.

On the table was a centerpiece of ferns on which were two crossed flags. At each place around the table were rush baskets of bonbons, tri-colored ribbons forming the handle of the baskets.

The second floor of the White House during the Benjamin Harrison Administration.

The menu included bouillon, beaten biscuit in the form of little chickens with outspread wings, cakes and ice creams. Hovering over the young guests were their mothers, nurses and the ladies of the White House.

After everyone had eaten all the ice cream and cake they could manage the President and his namesake led the guests back to the corridor where they all danced the Virginia Reel.

CLEAR SOUP

4 pounds of juicy beef	1 soup bunch (herb bouquet)
1 knuckle of veal	1 small pod of red pepper
2 small turnips	2 small white onions
2 small carrots	salt

6 quarts of water

Boil six hours, then strain through a sieve. Let stand over night, then skim off all the grease, put into a kettle to warm and add sherry or madeira wine to taste.

FISH CHOWDER

Cut a medium-size shad or white fish, three or four potatoes, one onion and a quarter of a pound of bacon into small pieces. Fry the bacon and onions a light brown. Put a layer of potatoes in a saucepan, over that a layer of fish, then a sprinkling of onions and bacon and a layer of tomatoes; season with pepper and salt, alternating the layers until all is in. Add enough water to cover, place over a moderate fire and let simmer twenty-five minutes. Boil one pint of milk, thickening it with cracker crumbs; let stand a moment and then add to the chowder. Now stir for the first time, let boil an instant, season if not strong enough to taste, and serve hot.

SAUSAGE ROLLS

Make a light biscuit dough (which uses milk) and let it rise overnight. In the morning roll it out thin and cut into shape with a biscuit cutter. In the center of each, place a roll of sausage the size of a good-sized hickory nut and roll it up in the dough. After letting this stand in the pan for a few minutes, bake and serve hot. (The rolls are also good cold, and when Mrs. Harrison was a child she took them to school for lunch in bad weather.)

The Harrisons, who brought morning prayer back to the White House, also brought frequent dancing. Mary McKee gave a ball in the East Room for young ladies and gentlemen of her acquaintance early in her father's administration. She also gave a final reception before he left.

The last event was held in the Red Parlor with only a few friends invited for between five and six o'clock. But the friends brought friends and they stayed on and on for nobody wanted to say good-bye. The President joined them and talked with several people that he had not seen for a year. Also there was Mrs. McKee's husband, who was soon to take the family back to Indiana.

Caroline Harrison, the President's wife, who met her husband when he was one of her father's students at Miami University, Oxford, Ohio, had hoped to get a new White House for the President when they first moved in. However, she got only a grant of $30,000 from Congress but with that she spruced up the old mansion. Electric lights and additional bathrooms were installed.

An artist who delighted in drawing orchids, she designed a new china service for the White House. But instead of putting the orchid, a symbol of luxury on the pieces, she designed a border of goldenrod and Indian corn with a second border of forty stars, and in the center of the plates, the shield and eagle of the United States.

Periods of mourning marked the Harrison years. His wife's father died, and then in the summer of 1892, in the midst of the political campaign, Caroline herself became ill with cancer and died in October. Her funeral was held in the East Room. Years later the President married her niece, Mary Dimmock.

Before his wife's illness, however, the White House was the scene of a joyous celebration when President Harrison, the grandson of an early President, was notified of his nomination for a second term. Members of the House and Senate were there to hear the nomination committee make its report, as were several hundred invited guests.

The President shook hands heartily, and the air was filled with joy. One guest stood on a satin chair and gave three cheers for the President. Cabinet members took turns passing lemonade and salad in the State Dining Room, while the President hurtled Baby McKee in the air in the hilarity. But it proved to be premature, for in November, Harrison was defeated by Grover Cleveland. Once again, the tenancy of the White House was changed and some new, old faces returned.

The Vermeil Collection, a group from the extensive silvergilt bequest to the White House of Mrs. Margaret Thompson Biddle in 1956. The collection, chiefly English and French, covers a wide span, from the Renaissance to the present.

The selection includes: top shelf, either end, Madeira cups, London, 1844; rear center, tray with pierced gallery, by Walker & Hall, Sheffield, 1825; front center, tureen and tray, festoon decoration by Philip Rundell, English, ca. 1820. Center shelf, left, footed cup engraved, marked by Robert, David and Samuel Hennell; center, French tureen ca. 1735; right, footed cup, Irish, probably Joseph Jackson, late 18th Century. Bottom shelf, left, coffee pot and hot water pot by Odiot, French, early 19th Century; center, monteith, or wineglass cooler, unidentified marks, ca. 1740; on either side, footed salt, acanthus decoration, London, Philip Rundell, early 19th Century; right, pitcher, by Paul Storr, London, ca. 1820.

William McKinley

chapter 15

SOCIAL CRISES AND CONQUESTS

President William McKinley's first administration, succeeding Grover Cleveland's, consisted of a series of social and national crises. He managed to overcome them all and win re-election to a second term, but that was to be cut short by an assassin's bullet.

He gave grander state dinners than did any of his predecessors, and the social responsibilities of the White House fell largely on his own shoulders and his staff. Ida McKinley, his beautiful, well-educated wife was an invalid, and could do little to carry out the social obligations of the First Lady. But her comfort and care came first with McKinley, and much of the social burden he bore was in her behalf.

The two were devoted and seldom apart. She was the daughter of a wealthy banker in Canton, Ohio, and was working as a cashier in his bank when she met McKinley, a young lawyer. After their marriage a series of calamities, including the death of her mother and the loss of two baby daughters within three years, shattered her health and left her an epileptic.

Even though excitement often brought on an attack, she insisted on going with him everywhere and participating in the social life at the White House.

The President modified the traditional seating at state dinners to have Ida at his side, and when a seizure came on, he threw a silk handkerchief over her face until she regained her composure. These attacks were called fainting spells, but she did not fall. She became rigid, with her head thrown back in a contorted expression.

At receptions Mrs. McKinley sat in a throne-like chair with the President at her side. She smiled and bowed as the guests passed in front of her and the President shook hands with the thousands who came.

Often cabinet wives stood with Mrs. McKinley and if a guest put out a hand to her, it was clasped by a cabinet wife who said "Let me shake your hand."

McKinley made many innovations in White House entertaining. He settled the long-debated question of who ranks first at social functions, the Vice President or ambassadors. He put the Vice President first, and this led to the practice still followed today of ranking the cabinet ahead of diplomats at official functions.

When the White House unwittingly gave an ambassador a seat of honor at a dinner attended by the Chief Justice, it caused a social storm. Chief Justice White took offense and when the company repaired to the East Room, he left without coffee, cordial, smoke or story, declaring he would demand a plan of the seating before accepting an invitation to dinner anywhere in the future.

The diplomatic corps also had its moments of social crisis at the White House. McKinley began the practice of giving each gentleman an envelope with the name of his dinner partner inside. At a dinner for the diplomatic corps, the Mexican Ambassador, Manuel de Aspirez, found the name of Madame Hengelmuller, wife of the Austrian minister in his envelope.

The Mexicans and Austrians were not speaking, because as a former Army lieutenant, the Mexican had been commanded to execute the ill-fated Austrian Emperor Maximilian, and thereafter no Austrian would call upon or speak to a Mexican in any capital. As a diplomat, however, Aspiroz handled the embarrassing moment with tact. He persuaded a Senator to accompany Madame Hengelmuller to dinner and he took the Senator's chair at the end of the table.

At another dinner for the cabinet, Admiral George Dewey, hero of Manila, and his wife did not receive seats of honor. Mrs. Dewey, a tiny woman from a wealthy and influential family, became indignant and the minute the guests arose from the dinner table, she dragged her admiral husband home.

The year before, however, on October 3, 1899, President McKinley had given a magnificent stag dinner in honor of Admiral Dewey following his return from the Philippines. Eighty guests were invited including representatives of the Supreme Court, cabinet, congress, army, navy, the governors of nine states, presidents of three universities, and the brother and son of the Admiral. It was one of the largest state dinners held in the White House up to that time, and the table was set in the first floor corridor, as the number exceeded the capacity of the State Dining Room.

The guests were received in the East Room. Then the President offered his arm to the Admiral and led the procession to the hall, which was decorated with palms, ferns and rubber plants.

The table itself, with thirty-seven guests on each side and three at each end, was profusely decorated with pink begonias and deep toned red roses, and lighted by hundreds of wax tapers under red and silver shades.

Music was furnished by the Marine Band which played selections saluting the returned hero. They included "Admiral Dewey," by Santelmann; "I Hail The Conquering Hero," by Handel; "Overture Jubilee," by Weber; "Voices Across the Sea," by Sousa; "Fantasie: Scotch Aires" and the national anthem.

He initiated a new method of sending out invitations. Instead of issuing one at the beginning of the year to cover all receptions, he sent separate invitations to each, and inaugurated the practice of having a number of young people present at all his functions to enliven them.

Young ladies found the White House a happy hunting ground for catching a husband. After one young heiress married a British earl, the other girls seemed more determined than ever to stalk down a diplomat in a foreign uniform and with a title. Often they were joined in the chase by doting mothers equally as eager for such a match. The diplomats had little chance to escape as the pretty girls hung on to their every word if not their arms.

Two thousand persons were invited to his first official reception on January 18, 1898 at which he and Mrs. McKinley received in the Blue Room under a canopy of smilax in which red, white and blue incandescent lamps were placed to give a patriotic glow.

A patriotic theme was used also at the reception for the Army and the Navy on February 11th. American flags were draped to form window curtains in the East Rooms, in festoons from the ceiling and over the doorways. Floral anchors and stars formed of roses were used in the decorations, and American flags hung all along the corridors interspersed with floral garlands above the doorway. At one end of the corridor hung the flag of the Secretary of the Navy, and at the other, that of the Secretary of War.

So lavish were the floral decorations that came from Government greenhouses, the flowers and plants would have cost two thousand five hundred dollars had they been purchased commercially.

President and Mrs. McKinley receive guests at a reception for the diplomatic corps.

The American flag was used in the summer of 1897 for wrapping a seventy-eight-pound prize watermelon, two and one-half feet long and six feet in circumference sent from Georgia as a gift to the President.

These were popular White House party desserts:

TUTTI FRUTTI

Line the interior of a three-part lemon-shaped mold with one pint of vanilla ice cream, cut four ounces of candied apricots into small pieces, also four ounces of candied cherries into halves. Mix these well together. Spread half the quantity of fruit over the cream in the mold and pour in one pint of raspberry water ice. Sprinkle the balance of the fruits on the water ice and fill the mold with one pint of pistachio cream ice. Cover the mold tightly, place in freezer for an hour and a half. Turn out on a glass stand and serve with the following sauce:

Put in a bowl one pint of whipped cream, two ounces of sugar and one gill of maraschino. Beat well together for two minutes. Pour over the tutti frutti and serve immediately.

CHANTILLY CHARLOTTE

Line a dome-shaped mold with ladyfingers, arranging them artistically. In a bowl whip one pint of thick cream. Add gradually four ounces of powdered sugar and one-half ounce of soaked gelatin. Flavor the cream with essence of violets. Pour cream carefully into the mold and set in freezer for an hour or two. When it is ready to serve, turn contents of mold onto a fancy dish and serve.

President McKinley inaugurated the practice of having the Marine Bugler announce the arrival of the President, rather than playing "Hail to the Chief" as had been done previously and is done today.

At their first state dinner, the McKinleys had sixty-two guests and placed the table in the long corridor with the Marine Band playing in the conservatory. As had been the practice for some years, each lady received a bouquet of roses at the table, and each gentleman a boutonniere of lilies-of-the-valley.

McKinley hired a plain cook for his family meals but for state dinners he hired a French chef from New York. Dinners seldom had more than twelve courses and often were limited to eight. However, the service was leisurely and the dinners sometimes lasted for two hours. Eventually the President and his wife arose and the ladies passed to the reception rooms and the men to the smoking room where coffee was served. (When dinner was served in the corridor, the State Dining Room was used as the smoking room.) After a brief period of informal conversation, the men joined the ladies in the reception rooms where tea was served.

Mrs. William McKinley

The custom then, as now, was that no guests left until those of highest rank had taken their leave. Out of consideration to others, these officials left immediately after tea had been served. By eleven o'clock, all the guests had departed.

Because of her frailty, Mrs. McKinley often did not eat the food that was prepared for the dinners, but ate a few crackers from a plate placed at her side.

At McKinley's receptions, a second military band often alternated with the Marine Band in providing music, and no refreshments were served, not even at the invitational affairs. One exception, was the reception for the Episcopal convention to which some two thousand were invited.

After receptions, however, the receiving parties which included members of the cabinet, high officials in government and persons invited to stand behind the line or with the President and First Lady in the receiving line were entertained at a supper in one of the upstairs rooms.

President McKinley enjoyed all entertaining but especially the public receptions, and said he got more from the friendly callers than they got from him, for they always had a smile and a cheerful word. Three times a week he held noon receptions for the public when he stood near the door and shook hands with the visitors. If Ida felt well enough, she sat at his side.

December 12, 1900 was "centennial day" in Washington, and the President held a morning reception for governors of the states and territories at which a model and drawings for an enlarged White House were displayed. Then the doors were opened to the public and McKinley set a record for handshaking. He clasped four thousand eight hundred and sixteen palms in less than two hours, an average of forty-six per minute.

The McKinleys gave many small dinners, after which he read or sat in a rocker on the South portico looking out over the Washington Monument. Mrs. McKinley whiled away her invalid hours crocheting or knitting. She made three thousand five hundred pairs of slippers that she gave to church bazaars during her years in the White House.

Because of their open friendliness, the McKinleys were popular and he won re-election in 1900. However, he was not to serve his term. In September of the following year while in Buffalo attending the Pan-American Exposition, he was shot at a public reception and died eight days later.

It was an age of tranquillity and prosperity; a time to evoke nostalgia. But with McKinley's administration, the United States entered the twentieth century and was coming of age as a world power.

A Saturday afternoon Marine Band Concert on the South Lawn.

Theodore Roosevelt

chapter 16

The Revolutionary Roosevelts

From a social point of view, there was never another administration like that of Theodore Roosevelt, who was sworn in as president on September 14, 1901, the day William McKinley died. Though only forty-two, he was already a national hero from his feats in the Spanish-American War. He was also a popular vice president.

As President, his popularity soared higher, and in this climate he and his forty-year-old second wife decided to have a social administration as outstanding as money and effort could make it.

They discarded old entertaining patterns and established new ones. There was scarcely a day that some type of social function was not held at the White House. If not by the President and First Lady, it was by or for one of their six exuberant children ranging from seventeen-year-old Alice, to four-year-old Quentin.

Mrs. Roosevelt inaugurated a series of public receptions on December 15, 1901, throwing the White House open from three to five o'clock for guests to call. She received them in the Blue Room, and more than one thousand women, mostly representative Washingtonians, came to see the new mistress of the mansion. The guests were presented by Colonel Theodore A. Bingham, a social aide. To obviate handshaking, Mrs. Roosevelt carried a large bouquet of white and purple orchids, but she greeted each guest with a smile and a word.

Two days later she was "at home" from four to six o'clock to two hundred invited guests, again receiving in the Blue Room, and on Christmas eve she entertained at a tea for four hundred invited to meet her sister, Miss Carew, who was spending the holidays at the White House.

The big event came, however, on January 3rd. It was Alice (called Princess Alice) Roosevelt's debut party and not since Dolley Madison's day had there been a social event to compare with it.

It was not all Alice had wanted, however, She had dreamed of a cotillion with champagne. What she had was a dance with fruit punch. But it was front page news the next day, the headlines reading: "Historic Mansion Sends Forth Its First Debutante."

President Roosevelt took no part in receiving guests on this occasion. They were welcomed in the Blue Room by Mrs. Roosevelt who presented them to her stepdaughter, Alice. Except for cabinet members and relatives, the guests were limited to young people, some seven hundred in all.

The Blue Room was decorated with palms and yards and yards of asparagus vine and smilax festooning the ceiling and entwining the chandelier. Adjoining rooms were decorated with masses of flowers, and in the East Room the carpet had been removed and the floor covered with a close linen crash heavily waxed for dancing.

Here the four pictures of George and Martha Washington, Abraham Lincoln and Thomas Jefferson were veiled in smilax caught back with baskets of red carnations. Round plaques of holly hung over each doorway. The mantels were banked with masses of cut flowers, the two west ones in green and yellow, and the others with two thousand roses, carnations, narcissus and hyacinths.

Dancing began at eleven o'clock with the Marine Band providing the music, selected especially to please the debutante. It included:

Mrs. Theodore Roosevelt

Waltz—"The Debutante," Santlemann.
Two-step—"Volia les Soldales," Eilenberg.
Waltz—"Therosen," Faust.
Polka—"The Bride Elect," Sousa.
Lancers—"The Strollers," Englander.
Waltz—"The Little Duchess," DeKoven.
Two-step—"Russe," Gasse.
Lancers—"The International," Moses.
Waltz—"Tales from Kiffahuser," Opitius.
Polka—"Tourists Souvenir," Bergenholls.
Waltz—"Wiener Blut," Strauss.
Lancers—"The Burgomasters," Luders.
Two-step—"Hop Along Sing," Armand.
Waltz—"Floradora," Stuart.
Polka—"Dem Goo Goo Eyes," Chattoway.
Waltz—"Sagamore Hill," Groebl.
Lancers—"The Foxy Quillet," DeKoven.
Waltz—"To Thee," Muelier.
Two-step—"Frangera," Costa.
Waltz—"The Beautiful Blue Danube," Strauss.

Alice wore a clinging gown of white chiffon built on taffeta, with elaborate garniture of tiny white rosebuds on the skirt and bodice.

Shortly after midnight, a buffet supper was served in the State Dining Room, decorated with red roses and poinsettas on the mantel and begonias and carnations on the table. The conservatories, where the west wing executive offices are today, were open and brilliantly lighted for promenading. One couple who went there to rest among the flowers stayed so long they suddenly realized the White House was quiet and went back to find the musicians packing their instruments and attendants putting out the lights.

Only the day before, President and Mrs. Roosevelt had held a public reception to which nine thousand visitors came. It was ideal weather that made waiting in line of little consequence to the crowd that gathered at ten o'clock in the morning. It was two thirty before the last person had gone through the receiving line in the Blue Room. Carriages and pedestrians were admitted at the West gate, while the East gate was used as an exit. A squad of mounted officers rode up and down Pennsylvania Avenue in front of the White House and patrolled the crowd.

It was arranged so that dignitaries would enter the Blue Room first, where the Roosevelts stood with cabinet members. The diplomatic corps was admitted first and then the door to the outer corridor closed. The result was that members of the Supreme Court and District Judges were kept waiting in the corridor. They were soon joined by late-arriving members of the diplomatic corps and then the outer door was closed. As a result, Senators and Representatives were left waiting on the porch while a large group of ladies who had been admitted earlier were received. Some members of Congress resented this, and went home in a huff.

It did not hurt Teddy Roosevelt's popularity with Congress, however. Shortly thereafter its members voted him more than half a million dollars for the first rebuilding of the mansion since 1815. A major change was the enlargement of the State Dining Room so that dinners for more than one hundred guests could be held there. This was made possible by incorporating into the room a former hall and stairway. In the rebuilding, the walls were paneled in English oak on which Theodore Roosevelt hung his animal heads and a handsome silver chandelier in the center of the ceiling. It matched the silver light sconces[5] on the wall.

The new mansion was opened for the social season beginning in December, 1902, but even before that the Roosevelts held state and special occasion dinners in the East Room or the first floor corridor.

[5] An ornamental wall bracket for holding a candle or other light.

Among the goodies were:

SAND TARTS

1 cup butter
2 cups sugar
2 whole eggs
1 egg, separated
2 teaspoons vanilla
4 cups sifted enriched flour

Cream butter to consistency of mayonnaise. Add sugar gradually, while continuing to cream, until light and fluffy. Add whole eggs, one at a time, beating well after each addition. Beat in egg yolk and vanilla. Stir in flour. Mix well.

For Sand Tarts: Roll out thin on lightly floured board. Cut with round cookie cutter 2½ inches in diameter. Beat remaining egg white slightly. Brush on cookies. Sprinkle with cinnamon-sugar. Bake on greased cookie sheets in moderate oven, 350 degrees, eight to ten minutes. Makes about six dozen cookies.

For Christmas Cookies: Cut in fancy shapes. Do not use egg white or cinnamon-sugar topping. Bake as above. When cool, decorate with tubes of decorating gel. Yield depends on size of cutters used.

CLOVE CAKE

½ cup butter
½ cup milk
½ cup molasses
2 cups flour
2 eggs
3 cups seedless raisins
1 teaspoon of soda (put in molasses)
½ teaspoon each of cloves, cinnamon and allspice
1½ teaspoons nutmeg

Mix and beat well. Bake in a greased mold.

One especially auspicious event held in the East Room was the dinner on February 24, 1902, for Prince Henry of Prussia, brother of the Emperor of Germany. Decorations consisted of thousands of tiny electric lights hung in the shape of stars, anchors, ropes, canopies and such, while the decorations on the large crescent-shaped table were pink-shaded candelabra and large vases of pink roses and primroses. Pink and white azaleas banked the mantels and ropes of smilax caught between the electric lights, hung from the ceiling and chandeliers. German and American flags were draped over the large window at one end of the room.

Each guest was presented a three-page program tied with red, white and blue ribbon. The cover was decorated with German and American flags, the German and American eagles and the emperor's yacht, Meteor. The Meteor had been built in a New Jersey navy yard, and was to be christened by young Alice Roosevelt before the Prince sailed home to Germany.

The dinner menu was as elaborate as the decorations. It consisted of:

Huitres sur Coquille
Croûtes panacheés
Potage Consomme Brunoisé
Olives — Celeri frise — Amandes salees
Terrapin a la Baltimore
Filet de Boeuf Hambourgeoise
Chapon a l'Ambassadriee
Petits pois — Sauce supreme
Asperges Sauce Mousseline
Punch
Sorbet Imperial
Canard Canvasback Pooti
Hominy — Salade de Saison
Glace
Petits Fours — Cerises Fondantes
Marrons glaces
Cafe

The wines included Marcobruner, '93; Sherry Amontillado; Moet & Chandon Brut Imperial; Chateau d'Arsac Grand vin Le Monteil 1893; Moet & Chandon White Seal; and Apollinaris Liquers. The punch was served in small boats flying the Meteor's flag.

It was a stag dinner and many of the guests came in uniforms bedecked with medals and ribbons of decorations.

The first social season after the remodeling of the mansion got off to an excellent start with a state dinner on December 18, 1902, in honor of the cabinet. Seventy-five guests were invited and for the first time they entered through the South front, walked through the lower level hall where hung the historic collection of portraits of former Presidents and First Ladies, then up the white marble steps to the main floor and the newly decorated state rooms.

The East Room, though not quite finished at this time, had white walls and heavy gold drapes at the windows, a color combination in effect today. The eight old marble mantels had been replaced by only four, all of colored marbles.

In the State Dining Room, the table was decorated in pink and white, with the Monroe mirror plateau in the center of the crescent-shaped table, and six bisque statuettes of mythological subjects placed among the roses surrounding the plateau. A corsage of bride's roses marked the place of each lady, and a single rose each gentleman's setting.

In 1904 President Roosevelt was presented a golden goblet about twelve inches high by the San Francisco Chamber of Commerce, and at every state dinner after that his place at the table was marked by that golden goblet.

Once a week Mrs. Roosevelt invited the wives of cabinet members to the White House, ostensibly to sew, but actually to discuss social events and develop ideas for making them run more smoothly. They were not adverse to a bit of gossiping or devising ways to curb social situations of which they disapproved.

The wife of one high official was having an affair with an attache from one of the Embassies. After discussing the matter, Mrs. Roosevelt sent the woman word that her White House invitations would be curtailed if she continued the romance. The woman chose to end it.

Edith Roosevelt gave a musicale nearly every Friday evening during the social season, and the concert and reception were usually preceded by a small dinner. Between two hundred and five hundred guests were invited to the musicales and the President's wife received them in the Green Room. Just as the concert began, the President joined the audience, seated on tiny gilt chairs in the East Room, and at the end he arose and thanked the performers. Ices and punch were served on trays in the East Room where the President and his wife mingled with guests.

At one musicale the artists were Edith Chapman, soprano, and Francis Rogers, baritone, with Bruno Huhn at the piano; at another the guests were invited to hear Paderewski, in whose honor a small dinner for fourteen guests preceded the concert.

The Roosevelts entertained often and elegantly but not extravagantly. During one six-month period they held one hundred eighty private entertainments. They restored the practice of serving refreshments at receptions, and instead of having state dinners catered as the Grants and others had often done, the Roosevelts employed three French chefs to prepare the food on such occasions. Mrs. Roosevelt was assisted in her planning and social duties by Bell Hagner, her social secretary, who provided her own footnote to White House social history.

At one private dance, Miss Hagner, wearing a red velvet gown with a long heavy train, was dancing with a diplomat. Suddenly there was a "thud" and guests turned to see Miss Hagner prone on the polished floor. Several rushed to pick her up and found her almost speechless with laughter. They discovered why when they lifted her to her feet. From underneath her train crawled the little diplomat with whom she had been dancing.

Typical of the official social schedule during the Roosevelt regime was this calendar which listed eight major events:

December 12—Cabinet Dinner
eight p.m.
January 1—Reception
eleven a.m. to one-thirty p.m.
January 3—Diplomatic reception
nine to ten-thirty p.m.
January 10—Diplomatic dinner
eight p.m.
January 17—Judicial reception
nine to ten-thirty p.m.
January 24—Supreme Court dinner
eight p.m.

January 31—Congressional reception
nine to ten-thirty p.m.
February 7—Army and Navy reception
nine to ten-thirty p.m.

Washington officials are jealous of their rank and position and take offense if not given the recognition they feel is due. Speaker of the House Joe Cannon expressed his resentment to Teddy Roosevelt that he was never a guest of honor at the White House. The following year, Roosevelt added a dinner for the Speaker of the House of Representatives to his social schedule.

Teddy Roosevelt's years in the White House were marked by two other singular social events. One was the wedding on February 17, 1906 of twenty-two-year-old "Princess Alice" Roosevelt to the much sought-after thirty-six-year-old Congressman Nicholas Longworth from Ohio.

The East Room was turned into a beautiful bridal setting for the ceremony and all the furniture removed from the other state rooms to make space for the nearly one thousand invited guests. Every room was banked with palms and flowers.

At noon, as the Marine Band played the magnificent "March from Tannhauser," Alice entered the East Room with her hand on her father's arm. There were no bridesmaids, maid of honor or flower girls to share the spotlight with the bride.

The Episcopal Bishop of Washington, the Right Reverend Henry Yates Satterless, performed the ceremony assisted by the Reverend Roland Cotton Smith.

After the ceremony, guests gathered in the State Dining Room for the wedding breakfast catered by Rauscher, the capital's leading caterer. The menu included croquettes, pates, salads, assorted sandwiches, ice cream in the shape of hearts, wedding bells and wedding rings, petit fours, champagne, claret punch, lemonade, coffee and tea, and the Bride's cake.

Another superior event was the debut party for Teddy Roosevelt's younger daughter, Ethel. As at Alice's wedding, all the state rooms were cleared of furniture to provide space for dancing. Supper was served in the rooms on the ground floor of the mansion.

Another memorable party was that held on Christmas Eve one year by Mrs. Roosevelt for six hundred children of officials in the administration. Instructions were sent out that adults could accompany only the most timid of the children, and the nurses were kept downstairs while the party was in progress.

Mrs. Roosevelt received the guests, who went to the East Room for a concert by a popular boys' musical group. Afterwards in the State Dining Room, the President helped serve the creamed oysters and other goodies selected for the youngsters.

The State Dining Room in 1902 when Theodore Roosevelt was host in the mansion.

William H. Taft

chapter 17

A Silver Wedding Anniversary

The social revolution begun at the White House by Teddy Roosevelt continued during the administration of his personally selected successor, portly, cheerful William Howard Taft, and his charming, friendly wife, Helen, of Ohio.

State dinners during the Taft years were large, elaborate and beautifully served. The table often glittered with crystal wine glasses full of bubbling wine although the President himself was not a drinking man.

He began the present day practice of having a musicale or some type of professional entertainment after state dinners. The performing artists, then as now, were not paid for their performances at the White House. However, Taft and his successor, Woodrow Wilson, presented them with gold medals costing about fifty dollars. In recent years they have been presented autographed photographs of the President and First Lady.

By far the most outstanding social event of the Taft Administration was their twenty-fifth wedding anniversary celebration on June 19, 1911.

The White House gardens and lawn were transformed into a fairyland and the state rooms became a bower of flowers. Electronics had touched the trees on the grounds with magic, and they blazed in red, white and blue bulbs. From atop the nearby Treasury Department building, a searchlight played on the American flag waving from the top of the mansion, and over the portico another flag of red, white and blue incandescents shimmered.

The fountain in the center of the grounds was the focus of another searchlight casting rainbow hues on the shooting waters. All around the outer edge of the eighteen acres of lawn were Chinese lanterns in various designs, barely six feet apart and gently swinging in the breeze. In the ancient elms and oaks were streamers of lights looped here and there among the branches. In pines and cedars, clusters of smaller bulbs were entwined around the trunks.

Some eight thousand persons were invited. More than five thousand came from both far and near.

The Marine Band, stationed on the South Grounds near the fountain, played the national anthem as the President appeared on the South balcony and then played the wedding march from Lohengrin as he descended the steps with Mrs. Taft on his arm. They walked together to the place on the lawn for the presentation of guests. There a canopy was formed by the branches of two trees joined by an electric sign flashing "1886-1911."

Preceding the President and First Lady were six presidential aides, and following them were members of the cabinet. The guests passed in two lines down winding walks to where the President and Mrs. Taft stood. As the lines moved, seventy members of the Washington "Saengerbund," led by Heinrich Hammer, sang songs.

Helen Taft wore a dress of white brocade embroidered with silver roses and carnations. The upper part of the gown was draped with a fichu of rose point lace, one of her wedding gifts, caught up on the left side with a single large pink rose. Her jewels were a tiara-like bandeau and dog collar of diamonds presented to her by the President.

There were chairs on the lawn for those who wanted to sit outside, or guests were permitted to wander through the state rooms where the Engineers Band played for dancing in the East Room. In the lower level rooms of the mansion were displayed the anniversary gifts including a seventeen hundred dollar silver service from the House of Representatives and silver compote dishes from the Senate.

Mrs. William Howard Taft

At eleven o'clock the State Dining Room was opened and a buffet supper served. Champagne was available as was wine punch. In the center of the table was a mammoth wedding cake baked by one of New York's leading caterers who had made the decorations a work of art. Twenty-five crystal hearts were embedded in scrolls at intervals on the sides. Alternating with the hearts were twenty-five miniature silken reproductions of the "stars and stripes" and the President's flag. On top was a cornucopia filled with sugary reproductions of the gardener's art and dainty cherubs. Around the base were roses cut from their stems and flung against the side of the bottom tier. Fluttering on the edge of the cake were turtle doves made of sugar.

Tables and chairs were set up on the west terrace for those who wanted to take their food outside and watch the artistic play of lights on the grounds below as they ate. The President and Mrs. Taft sat on the South balcony after the ceremonies were over and here they again received intimate friends and relatives.

Watching the festivities from the sidewalks was a crowd estimated at fifteen thousand and the following night the grounds, still decorated, were open to the public while the Marine Band played an outdoor concert for their enjoyment.

A diplomatic reception on March 12, 1909, was the Tafts' first official entertaining and they amazed guests with their friendliness and informality. Instead of leaving after all the guests had been presented, they stayed and mingled with them, talking informally. The guests, accustomed to having the President and First Lady leave first, didn't know when they should excuse themselves. Finally, Mrs. Taft had an aide announce that guests could leave when they wished without any adieus.

The menu for the reception included lobster à la Newburg, chicken pates, salad, assorted sandwiches, rolls, cakes, ice cream, candies, coffee and punch.

The Tafts always served supper at their receptions (both the public and invitational events), and sometimes there was a mad crush of guests near the buffet tables. Sometimes, too, guests wanted to take home souvenirs from the table and Secret Service agents were stationed at the doors to stop them. The agents would tactfully say to those whom they had seen take a souvenir, "Let me put that spoon down for you." Or, "Won't you let me dispose of that table napkin for you. I'm sure it is in your way."

President Taft often used his dinners for winning support of influential members of Congress for his legislative proposals. On occasions, he served two kinds of wine at dinner, the vintage wines going to guests whose support he was seeking; the inferior brands going to the other guests.

During their first two months in the White House the Tafts gave two garden parties, six major dinners,

A place setting for a state dinner in the 1960s shows napkins with the seal of the United States woven into the fabric.

several smaller ones, and had entertained every Senator and most of the important members of the House of Representatives at least once. The social strain had its impact on Mrs. Taft, who suffered a mild stroke and was forced to curtail her activities.

She hired an executive housekeeper, Elizabeth Jaffray, to replace the steward her predecessors had employed, and not only had her help in keeping the domestic side of the White House functioning smoothly, but her assistance in planning menus and various entertainments. Often the same basic menu was used again and again with slight variations to conform to the occasion.

At a dinner dance for the Tafts' debutante daughter, Helen, on February 11, 1911, the menu read:

Anchovies
Clear Soup
Filet of Bass with Oyster Crab Sauce
Mousse of Chicken—Mushroom Sauce
Filet of Beef   Truffles
Stuffed Tomatoes with Mushroom Sauce
Asparagus
Pheasants   Orange Salad
Macaroon Ice Cream
Cake
Coffee
Champagne   Macaroons   Peppermint Paste

Less than a month later, March 4, 1911, the President gave a dinner for his 1878 class at Yale. The menu read:

Anchovies
Clear Soup   Custards
Filets of Bass with Oyster Sauce
Artichokes with Mushrooms
Saddle of Mutton   Mint Jelly
New Peas   New Potatoes
Patè de Fois Gras in Aspic
Duck   Mixed Salad
Cake
Coffee
Macaroons   Cherries in Candy   Peppermint Paste
Marshmallow Caramels

There was rarely a day the Tafts did not have guests for breakfast, luncheon or dinner, and often at all three meals. Sometimes the guests were invited on short notice, but Mrs. Taft learned to cope with her husband's impulsive invitations. She was also experienced in official entertaining when she became First Lady, having shared her husband's public life when he was Governor of the Philippines and subsequently Secretary of War.

Despite his weight, (over three hundred pounds), President Taft was light on his feet and loved to dance. So did Mrs. Taft. They organized a dancing class with a dozen friends and the White House aides, and met once a week in the White House for dancing to music on records.

On December 1, 1910, Helen Taft made her debut at a party memorable in the annals of White House entertaining. The East Room was decorated almost entirely of flowers that were sent to her and she held a bunch of lilies of the valley during the reception. Mrs. Taft greeted the twelve hundred guests, presented Helen, and then the President received them.

Mrs. Taft wore a blue chiffon gown, and her daughter, white satin with an overdress of pink chiffon with a shallow yoke of white lace.

The State Dining Room was decorated with begonias, and maidenhair fern banked the mantels. Pink rosebuds and pink tapers added beauty to the long table.

On December 30th a ball was given in honor of the debutante with three hundred guests, mostly young people her age, invited. The East Room was cleared for dancing, and an enclosure for the Marine Band built on the East Terrace outside the triple windows from which the glass panes had been removed so the music came through clear and smooth.

In January, 1913, two months before they left the White House, President and Mrs. Taft gave a romantic dinner for Mrs. Grover Cleveland in the mansion where, twenty-six years earlier, she had been married. Four members of Cleveland's cabinet and widows of two others were guests, as was the second wife of President Benjamin Harrison, now a widow. Also there was Mrs. Cleveland's fiance, Thomas J. Preston, an archeologist at Princeton, whom she married shortly thereafter.

Covers were laid for fifty-two guests and the table was decorated with one of Mrs. Cleveland's favorite flowers, jonquils, and maidenhair fern.

Mrs. Taft had a great sense of history and often used the china service of preceding presidents at small luncheons and dinners and found them an inspiration to lively conversation. She also used the West terrace outside the State Dining Room for dinners in hot weather because the White House was not yet air-conditioned.

The Tafts produced an eminent son, Robert, who was to perpetuate the family name in public service and narrowly miss becoming an occupant of the White House in his own right.

White feverfew, red and maroon Sweet William in a vermeil tureen on a console in the State Dining Room.

Woodrow Wilson

chapter 18

A SERIES OF WEDDINGS

The tempo of social and official entertaining during the Roosevelt and Taft administrations slowed to a leisurely pace when the scholarly, former college president, Woodrow Wilson, became President in 1913, and then ground almost to a halt when the United States entered World War I four years later.

President Wilson's eight years in the executive mansion were marked by one death and four weddings in the family, a record of nuptials for any administration up to that time. Three of the weddings were held in the mansion but the fourth (when the President, who became a widower in 1914, took a second wife), was held at the bride's home.

The other three weddings were those of two of Wilson's three daughters and of his niece.

President Wilson's first wife, the former Ellen Louisa Axson, daughter of a Presbyterian minister of Rome,

Georgia, was as interested in social reforms and improvement of the slums of Washington as she was in social life in the White House. So she portioned her time to serve both of these interests and her own family which, of course, came first.

She held receptions twice a week, often after having spent part of the day in the slums to see firsthand what was needed and to formulate ways to meet these needs. She was also a gracious hostess at the official dinners and receptions she and the President gave.

Mrs. Wilson brought her own cook from New Jersey to the White House to prepare the family meals, but she also kept the Swedish cook which the Tafts had hired through a New York employment agency to prepare the food for state and official functions. She also retained Elizabeth Jaffray, who had kept things humming smoothly as executive housekeeper in the preceding administration.

Mrs. Jaffray, who had assisted in planning the menus for the highly successful dinners during the Taft administration lent the same touch to the Wilson's dinners. The menu for the first diplomatic dinner given by the Wilsons was an example. It included:

Caviar in Ice
Turtle Soup with Cheese Straws
Mousse of Halibut with Lobster Sauce
Artichokes with Mushrooms
Filet of Beef
Stuffed Tomatoes String Beans New Potatoes
Asparagus with Hollandaise Sauce
Mallard Duck
Mixed Salad
Ice Cream Cake Macaroons Caramels Mints
Coffee

After a state dinner, President Wilson, who enjoyed mimicry, often entertained his family privately with impersonations of the guests, particularly the ladies, they entertained. He also liked to sing, and entertained his family and close friends by bursting into song at family gatherings.

The social highlight of the Wilson years was the elaborate wedding given his middle daughter, Jessie, the first to wed. She was married in a late afternoon ceremony on November 25, 1913, in the East Room before an audience of the Nation's top dignitaries and foreign envoys in Washington. The bridegroom was Frank Sayre, a young lawyer whom she had met when her father was Governor of New Jersey.

For the event the East Room was turned into a fairyland of flowers. A ribbon-bordered aisle led to a satin-covered prie-dieu on a white vicuna rug in front of the East window. Almost obliterating the window from view was a bank of flowers and greenery with tall candelabra on each side of the improvised altar.

As the Marine Band played Lohengrin's "Wedding March," the bridal procession entered, the tall, statuesque Jessie in a beautiful white satin gown trimmed with rare old point lace. The President, looking handsome and dignified, wore a dark gray cutaway. Margaret, her oldest sister and a concert singer, was a bridesmaid, as was her youngest sister, Eleanor.

After the ceremony a wedding supper was served in the dining room, and when the wedding cake was cut, it was Margaret who got the piece in which was hidden the wedding ring signifying she would be the next to wed.

Within six months invitations were out to another White House wedding but Margaret was not the bride. It was her youngest sister, Eleanor, who had become engaged shortly after Jessie's wedding, to an attractive widower, William Gibbs McAdoo, the Secretary of the Treasury.

Her wedding was smaller and less elaborate than Jessie's had been, because by this time the girls' mother had become ill and was not up to planning a big event. Only relatives, members of the Cabinet, and their wives were invited to the ceremony held in the Blue Room. However, all the state rooms were decorated with flowers for the event.

The East Room, through which the guests entered, was decorated with great clusters of Annuciation lilies and boughs of blossoming dogwood, for it was a spring wedding, on May 7, 1914. The Green Room was fragrant with lilies and roses, and in the Red Room were American Beauty roses and dogwood. Dogwood boughs also filled the alcoves of the long corridor, and pink roses brightened the oak-paneled State Dining Room.

Delicate white apple blossoms and white lilies decorated the Blue Room where a dais platform stood between the two bay windows overlooking the South lawn. The dais was covered with blue satin to match the wall, and a white vicuna rug laid over the top. On this rested the white satin prie-dieu tied with white ribbons and white roses and lilies.

After the ceremony the bridal couple received congratulations in the Red Room while the Marine Band played music in the hall. Then supper was served at small tables in the State Dining Room. Just as the guests were sitting down, one of the bridegroom's three little girls (of his first marriage) burst into tears, became hysterical and was taken from the room. The other guests continued the festivities and the menu for that wedding supper included:

Bouillon
Chicken Paté
Boned Capon with Peas
Virginia Ham Celery Salad
Strawberry Ice Cream
Cake
Coffee
Champagne

Three months later, on August 6th, Mrs. Wilson died at the age of sixty-four from a kidney ailment. The White House was plunged into mourning, and official entertaining was suspended for a period.

By spring, however, the grieved President had a new interest in life besides his burdensome job. He had met the attractive widow, Edith Bolling Galt, and was pursuing her in a secret courtship that caused gossip (as he predicted it would).

On December 15, 1915, he married her and brought a new First Lady to the White House. She made her first appearance as hostess at a diplomatic reception in January at which she greeted more than three thousand guests. She had never been to an official White House party before, but she played her new role well.

There followed a series of receptions, teas and luncheons. And when spring came, informal lawn parties were given in the White House gardens. The Wilsons received their guests under a green canopy, and refreshments were served at a large colored tent erected on the lawn.

Entertaining ground to a halt again as Wilson became engrossed in his campaign for re-election and soon thereafter the United States was plunged into war. So official entertaining was not resumed.

There were, however, many dinners and luncheons for allied leaders who came to Washington to confer with the President. One of these was a luncheon for Arthur Balfour, head of the British delegation. Mrs. Wilson and Helen Bones, the President's cousin who lived at the White House and was responsible for his meeting Mrs. Galt, were the only ladies at the dinner for fifty male guests.

The menu, selected especially for the men, consisted of:

Fruit Cocktail
Clear Soup
Filet of Sole with Tartar Sauce
Filet of Beef with Mushrooms
Breast of Chicken
Hearts of Lettuce
Strawberry Ice Cream
Cake
White Rock

Another of these events was the dinner on August 23, 1917, for the Japanese Mission. Mrs. Wilson was the only lady at this event for seventy-eight guests. The table was decorated with scarlet and yellow gladioli and the menu included:

Muskmelons
Clear Soup — Vegetables
Roast Capon — Potatoes
Virginia Ham
Waldorf Salad
Peach Ice Cream — Cake
Coffee
Champagne, claret, sherry, scotch, whiskey, brandy.

President Wilson gave another wedding at the White House on August 7, 1918. This one was for a favorite niece, the daughter of his younger brother. The bride, Alice Wilson, of Baltimore, Maryland, was married to Isaac Stuart McElroy, Jr., a Presbyterian minister, in the Blue Room at the end of a hot summer day—the hottest summer day the Weather Bureau had recorded in forty years.

But a storm broke just at six o'clock and cooled the air before the eight o'clock ceremony.

Again the Marine Band played the traditional Lohengrin and Mendelssohn wedding marches as the bridal procession came down the stairs. Again the little white satin prie-dieu was placed in front of an altar of flowers—white phlox and white hydrangeas gathered from the White House greenhouses. Again the State Dining Room was decked with pink roses and the table set for a wedding supper. But this time there were only sixteen guests, all members of the family.

After the war was over the President and Mrs. Wilson journeyed abroad to attend the peace conference. When they returned he traveled throughout the United States in a determined effort to win support for his League of Nations as an instrument of maintaining peace. There was little time left for official entertaining in Washington.

However, in August, 1919, he and Mrs. Wilson gave a large garden party for the wounded veterans at area hospitals and the President himself served ice cream and cake to the boys. This event became the forerunner to the annual garden party receptions held by almost every succeeding President for veterans and servicemen in area military hospitals.

President and Mrs. Wilson gave several other large receptions that year and at one of them a great many guests decided they wanted the same thing as a souvenir from the White House. At the end of the party four hundred of the heavy silver-plated spoons with "The White House" stamped on them, were missing. The solution to the problem of the missing spoons thereafter was to use spoons that were unmarked.

Before leaving the White House in March 1920, the Wilsons purchased a seventeen hundred piece china service for the mansion with an appropriation provided by Congress. It was the first American-made china purchased for the White House and came from the Lenox Company in Trenton, N.J. Designed by the firm's artist, the china had the presidential seal as the motif, with a gold rim outside a deep ivory border. The service plates had a rim of deep blue with a broad gold rim.

President Wilson became ill in September, 1919, and his loyal wife carried on for him, keeping not only the public away from the White House, but members of his own staff away from his sickroom. She served as liaison between his sick room and his cabinet as well as the public and this period became known as "Mrs. Wilson's Regency." There was no White House entertaining and the gates were closed again, just as they were in wartime.

Following are some of Mrs. Wilson's recipes, many of which are in her own handwriting now in the Woodrow Wilson House, 2340 S Street, N.W., Washington.

CLAM DIP

1 small can clams
1 package cream cheese

Drain off juice, add onion (grated), salt and pepper.

ANGEL ON HORSEBACK

Small raw oysters
Bacon
Tartar sauce
Bread slices in oval shape

Wrap oysters in three-inch strips of bacon. Fasten with toothpicks. Broil in oven until bacon is crisp. Remove toothpicks and serve on toasted bread ovals spread with tartar sauce.

The Marine Band plays from its traditional station in the entrance foyer for a White House party.

FRENCH DRESSING

1 can Campbell's tomato soup
1 cup tarragon vinegar
in which dissolve
1 tablespoon sugar
1 tablespoon salt

Cut up one onion and put all ingredients in quart jar. Fill jar with olive oil and shake thoroughly. Close tight and keep in refrigerator until used.

MRS. WILSON'S STUFFED LOBSTER

Boil or steam lobster in one cup vinegar, two cups of water. Cover tight and cook for twenty minutes; let cool still covered.

Make a thick cream sauce with only a pinch of tarragon for seasoning. Pinch the lobster into pieces (never cut it). Put in sauce, let stand over night. Put in shells when ready to serve.

HOT PEPPERED NUTS

Into an iron skillet over a low flame put a pound of paper-shell pecans, halved. Add two or three lumps of butter the size of walnuts. Stir frequently and when almost toasted add salt and cayenne pepper generously. Place in a warming oven to keep hot until served.

BAKED EGGS IN TOMATO SAUCE

3 cups tomato juice
6 hard-boiled eggs
1 small onion
1 clove garlic
1 small bay leaf
3 cloves
3 tablespoons butter
3 tablespoons flour
crumbs
½ cup grated American cheese
Salt and pepper to taste

Cook the tomato juice with the chopped onion, garlic, bay leaf and cloves for ten minutes. Blend flour and butter and strain over this the tomato juice. Stir well and cook until thickened.

Cut the eggs in half and place in a buttered baking dish. Pour over them the tomato sauce and cover the top with the bread crumbs and cheese that have been well mixed. Bake fifteen minutes in a moderate oven, or until the sauce begins to bubble through the crust.

LOBSTER CUTLETS

1 pint of lobster meat cut fine

Season with one teaspoon salt, one teaspoon of mustard and a little cayenne. Moisten with one cup of thick cream sauce. Spread this half an inch thick on a platter. Cut into shape of cutlets. Roll in crumbs, egg, and crumbs again. Fry in smoking hot fat. Drain and serve with a claw to represent the bone.

CREAM SAUCE

1 pint hot cream or milk
2 even tablespoons butter
4 heaping tablespoons flour
½ teaspoon salt
½ saltspoon pepper
Speck of cayenne

PEARS IN COINTREAU WITH FROZEN CREAM

Peel twelve perfect, firm Bartlett pears, leaving them whole with their stems. Put immediately in cold water with a few drops of lemon juice so they won't turn dark. In the meantime, make a syrup of four cups of light brown sugar and six cups of water. Boil for a few minutes, then put the pears in and cook until tender and transparent, but not mushy.

Fill a teacup with Cointreau, and when the pears are done, lift each separately and dip in the Cointreau until well saturated. Put them in a big glass dish, piling in a pyramid if possible. Continue to boil the juice until moderately thick, then add the Cointreau in which the pears were dipped and pour gradually over the pears. Place in the refrigerator to chill.

An hour and a half before dinner is announced, empty a pint of thick cream into one of the trays of the refrigerator. When ready to serve the pears, this cream should be just crystallized and not quite stiff. Put it into a very cold silver bowl and pass with pears.

Miss Jessie Wilson's (Mrs. Francis B. Sayre)

OATMEAL COOKIES

1 cup sugar
½ cup butter
2 eggs
2½ cups rolled oats
1 teaspoon baking powder
1 teaspoon vanilla

Mix entirely with hands. Cream butter, sugar and add oatmeal gradually. Then take spoon and add eggs, vanilla and baking powder. Drop from teaspoon on waxed paper and bake until brown in hot oven.

A view of the Red Room today showing the graceful Empire sofa that once belonged to Nellie Custis Lewis, granddaughter of Martha Washington.

Warren G. Harding

chapter 19

The Gates Open Again

When handsome, gregarious Warren Gamaliel Harding of Ohio, and his charming wife, Florence, moved into the White House in March 1921, it was as if a shade drawn over the mansion during President Wilson's illness had been lifted, and the public could see again what went on inside, and could go in and out again.

Mrs. Harding, a forceful, dynamic person who was in a large measure responsible for her husband's business and political success, assured the voters that they were just plain folks and intended to remain that way in the White House.

With great verve and enthusiasm she entertained at a series of garden parties attended by throngs of Washington officials and socialites who put on their gayest dresses and most glamorous hats to go to the White House again. Lemonade, punch, cookies, cakes and sandwiches were served from long tables beneath gay striped awnings. The Marine Band played and the Hardings stood under a huge oak tree to receive the guests. When it came time for them to go home, the Band played "The End of a Perfect Day."

Sometimes Mrs. Harding sent engraved cards to three thousand guests for her garden parties and never seemed to tire in shaking the hands of all who came. She frequently invited veterans at area hospitals to these garden parties and always wore what she termed her "old hat" so they could spot her in the crowd.

She and the President were both musicians and there was a lot of music at their parties. They both loved to entertain also and gave many dinners, receptions and luncheons. They invited their personal friends to small private dinners and rarely ever dined alone in the evenings.

Despite the warmth and graciousness of the Hardings' hospitality, however, their social events lacked the lustre of some of the previous administrations. One exception, though, was the social season of 1921-22 when the President invited the leaders of nine nations to Washington for a Conference on the Limitation of Armaments. The guests were entertained at the White House, and at the New Year's Day reception in 1922, the foreign delegations were guests of honor.

Mrs. Warren G. Harding

For this event the mansion was turned into a tropical garden with pink rosebushes and ferns banked on the staircase, palms and leaves in the East Room, and pink carnations and bouquets of garden flowers in the Blue Room.

Wines disappeared from the table at state dinners during the Harding Administration for prohibition had become law. However, the law was ignored at Harding's small stag parties upstairs in the family's private quarters. He often invited a dozen or more of his cronies to his quarters, where they played poker nearly all night. Mrs. Harding served as bartender and kept the glasses filled with the legally forbidden drinks. Food at these private parties rarely ever varied. It was always knockwurst and sauerkraut.

One of her typically American recipes was for:

MRS. WARREN G. HARDING'S WAFFLES

2 eggs, separated
2 tablespoons sugar
1 teaspoon salt
2 tablespoons butter, melted
2 cups milk
3 cups flour
3 teaspoons baking powder

Beat the egg yolks and stir in the sugar, salt and melted butter. Stir in the milk and enough of the flour to make a thin batter. Just before baking stir in the baking powder and fold in the egg whites, stiffly beaten. Bake in a hot waffle iron.

At state dinners given by President Harding, the table was usually decorated with large bowls filled with fruit and ripe Malaga grapes. The Wilson china was used.

In September 1922, Mrs. Harding became ill and went into seclusion for six months putting a damper on social entertaining that season. Eleven months later, while her husband was on a speaking tour to the West and Alaska, he became ill. He died on August 2nd, 1923, in San Francisco as his wife sat reading to him. The funeral was held in the East Room and a week later Mrs. Harding moved out of the White House.

Vice President Calvin Coolidge, a former Governor of Massachusetts, was sworn in as President in the middle of the night by the light of a kerosene lamp in the parlor of his father's farmhouse in Vermont. A few days later he and his wife, Grace Goodhue Coolidge, quietly moved into the White House from the hotel suite they had occupied during his three years as vice president.

Grace Coolidge, forty-four, friendly, pretty and in good health, was suited for the role of First Lady although she had little experience in official entertaining. There had been no official residence when her husband was Governor of Massachusetts, and they had done no official entertaining in their hotel suite when he was vice president. But she plunged into her new duties with common sense and enthusiasm, never forgetting that her first obligations were to her husband and their two sons.

It was not long until it was very evident that new tenants were in the executive mansion. Coolidge took a personal interest in the domestic details, initialed all household bills, approved the menus for official and family dinners, checked the table before the guests arrived, and even passed approval on the gowns worn by Mrs. Coolidge.

He was strict about the formalities of the office he held and insisted that his wife and sons be in full evening attire when they came to dinner, even when the family dined alone.

Despite his New England thrift and economy, his was the most expensive administration up to that time when it came to entertaining for Harding had gotten Congress to pass a law providing an allowance of $25,000 annually for official entertaining in the White House and for presidential travel.

Perhaps as a gesture of appreication or to improve his own relations with Congress, President Coolidge initiated a series of "alphabetical breakfasts" to which he invited a dozen or more members of Congress to the White House at a time for an eight o'clock breakfast. These social events became the talk of Capitol Hill and the target of many jokes.

The invitation went out by telephone the night before and the guests who couldn't find an excuse to avoid such an early morning engagement, arrived to

Pieces from the silver service purchased by President Jackson from the Russian Baron de Tuyll include hot-water pot, coffeepot, cream pitcher, vegetable dish, and wine bucket. Marked by a leading French goldsmith of the early 19th century, Martin Biennais, the silver was purchased in 1833 for $4,308.82.

be ushered into the Red Room where a wood fire crackled on the hearth. After waiting a few moments, the President came in, shook hands with each guest, and led them to the family dining room where the menu was the same at each breakfast: grapefruit, bacon and eggs, corn muffins, New England sausage, buckwheat cakes, Vermont syrup, toast and coffee.

One recipe used for the breakfasts was for:

MRS. COOLIDGE'S CORN MEAL MUFFINS

2 cups corn meal
1 cup flour
1 cup sweet milk
2 eggs well beaten
½ cup sugar
2 tablespoons baking powder

Mix ingredients together and bake in moderate oven until brown. Makes fourteen muffins.

The President, nicknamed "Silent Cal," did very little talking but ate in silence as those around the table carried on conversations with each other. Once the President poured his coffee into a saucer, added cream and sugar as the guest waited expectantly for him to drink it. He surprised them when he set the saucer on the floor for his dog, who had unusual White House privileges. The dog's drinking dish was kept on the hearth in the State Dining Room and more than one guest saw it there among the ferns during official functions.

Although Coolidge didn't have to pay for the refreshments and food he served his guests, he was thrifty about what he served. At his receptions he served nothing—not even ice water unless a guest asked for it.

At his first diplomatic reception one of the guests—a Mrs. William Vanderbilt, stood in line for a long time as did everyone else to be received by the President and First Lady. She turned to a young man beside her, from the Canadian Legation, and asked him to get her a drink of water as the heat of the crowded room was oppressive. The man returned with a paper cup filled with water from the cooler in the usher's office. He explained that the President had ordered that glasses and silver pitchers not be set out as it would encourage people to loiter and waste time. He expected the guests to shake hands with him and go home.

It seemed to White House staffers that no other President they knew about had as many idiosyncrasies as Coolidge and many are the stories told about them by the people who worked with and for him.

Elizabeth Jaffray, executive housekeeper for a while during the Coolidge administration, told of his looking in the kitchen one evening before a dinner for sixty guests, and seeing six Virginia hams ready to be served complained "I don't see why we have to have six hams for one dinner. It seems an awful lot of ham to me."

Mrs. Jaffray explained "But Mr. President, there will be sixty people here. Virginia hams are small and we cannot possibly serve more than ten people with one ham and be sure of having enough."

"Well, six hams look an awful lot to me," the President repeated.

SPINACH TIMBALES

1 cup cooked spinach
½ teaspoon salt
1 teaspoon onion juice
1¼ cups milk
2 eggs, well-beaten
2 tablespoons shortening
dash of pepper

Chop spinach. Add seasonings, onion juice, well-beaten eggs, milk and melted shortening. Stir just enough to blend. Turn into greased molds or custard cups, set in a pan of hot water which has been lined with several thicknesses of brown paper and bake for thirty minutes at 350 degrees F., or until a clean dry knife inserted into the timbales comes out clean. Unmold and serve at once.

A dinner for the President of Cuba on a Saturday night, April 23rd, 1927, was fraught with mishaps, none the fault of the President, however.

Covers were laid for twenty-six guests and of that number three had mishaps: one Cuban lost his necktie after it became unhooked during the dinner. An American guest broke his front tooth while eating. And another Cuban while seated at dinner, pushed through the bottom of the cane seat of his chair and when he got up at the end of the dinner, the chair hung from him.

All the while, a fire in the chimney was causing excitement outside. The fire was extinguished but men were placed on the roof to watch and control it in case it blazed up again.

At another state dinner, the guests had all arrived except one, and the President and First Lady were awaiting his arrival before making their appearance in the Blue Room. At nine minutes past eight, he arrived very obviously having been in violation of the eighteenth amendment. President and Mrs. Coolidge made their appearance while uneasy military eyes kept a close watch on the inebriated diplomat to see that he didn't get out of line.

Another night at a musicale, one of four given each season by Mrs. Coolidge, a minor government official came late and roaring drunk. He seated himself beside a friend, also intoxicated, on the back row of chairs set up in the East Room. There they carried on an audible conversation with much boistrous laughter and knee slapping during the musicale. Finally, the chief usher and two aides led them out of the White House and put them into a vehicle which removed them from the premises. Needless to say, they were not invited back.

When "Silent Cal" was silent he was often preoccupied and sometimes violated the formalities he so strictly expected of others. At a cabinet dinner on November 8th, 1928, two days after the election of Herbert Hoover, he left the table abruptly as soon as he had eaten his ice cream although candy, fruit and bonbons had not been served.

Everyone was embarrassed but none more so than Mrs. Coolidge who tried to remedy the breach by serving the candy in the parlors.

Despite Coolidge's idiosyncrasies and economies, Mrs. Coolidge maintained an atmosphere of hospitality and graciousness. She and the President gave four state dinners and four musicales each season plus a great number of receptions and teas. The Red Room was her favorite for holding teas. Each Monday and Friday she was hostess there to from twenty to thirty guests. Tea was served in front of the fireplace and it was a cozy atmosphere for intimate conversations.

Some of Mrs. Coolidge's recipes for desserts served in the White House are:

GRACE COOLIDGE'S LEMON PIE

1 cup sugar
1½ tablespoons flour
Juice of 1 lemon
2 eggs, separated
1 tablespoon butter, melted
1 cup milk

Combine the sugar and flour and stir in the lemon juice, beaten egg yolks and butter. Stir in the milk and fold in the egg whites, stiffly beaten. Line a pie plate with rich pie dough and bake it in a hot oven (450 degrees) for about ten minutes, or until the crust is set. Add the filling and bake in a slow oven (325 degrees) for about 35 minutes, or until the filling is set.

CUSTARD PIE

¾ cup sugar
2 eggs, beaten
1 tablespoon flour
2½ cups milk

Mix sugar and flour; add a pinch of salt; bake in a single crust; dust a little nutmeg on top after pie is baked.

COFFEE SOUFFLE

1½ cups coffee
1 tablespoon gelatin
⅓ cup milk
3 eggs
¼ teaspoon salt
⅓ cup sugar

Mix coffee, gelatin, and milk and heat in a double boiler. Add the yolks of the eggs, slightly beaten and mixed with ⅓ cup sugar and ¼ teaspoon salt. Cook until it thickens. Remove from the heat and add the whites of eggs beaten stiff and ½ teaspoon of vanilla. Mold, chill and serve with whipped cream. Serves six.

At one of their musicales, President and Mrs. Coolidge presented a small string orchestra directed by

Dr. Thaddeus Rich, which played sixteenth and seventeenth century music on rare old Italian violins, cellos, violas and double basses borrowed from the Rodman Wannamaker Collection.

The program was modeled after those given by the private orchestras of the court of those centuries and it was the first time some of the musical manuscripts had been played for an audience in two hundred years. One was a work by Domenicio Scarlatti, the others were compositions by Thomas Tompkins and William Byrd, composers of the sixteenth century.

Even "Silent Cal" could unbend and join in the fun when he wanted to. And he did in 1924 when he was seeking election to a full term of his own in the White House. He entertained a group of theatrical stars representing the Coolidge Non-Partisan League at a breakfast at the White House. In the group, which went directly to the mansion from Union Station, were Al Jolson, John Drew, the Dolly Sisters, Raymond Hitchcock, Charlotte Greenwood, Stella and Francine Larrimore, Gerald Griffen, Justine Johnson, Cecil Lane and others.

They gathered on the White House lawn with the President and Jolson, serving as spokesman, told the President that the theater profession was satisfied with his administration and was going to work to keep him in the White House. The usually stone-faced President was all smiles as Jolson led the group in singing their campaign song, "Keep Coolidge". . . .

"The race is now begun
And Coolidge is the one,
The one to fill the presidential chair.
Without a lot of fuss, he did a lot for us
Let's reciprocate and keep him there.
Keep Coolidge, keep Coolidge, for he will right all wrongs
He's never asleep, still waters run deep,
Keep Coolidge, keep Coolidge, he's right where he belongs.

Mrs. Coolidge joined in the singing and after the song was sung twice, the President invited the group to the State Dining Room for one of his New England breakfasts. Coolidge stayed until ten-thirty when he suddenly remembered he had a ten o'clock cabinet meeting. He was late in keeping that appointment.

Christmas was a festive occasion for the Coolidges who celebrated in New England fashion. A huge tree was decorated in the Blue Room, and one year Mrs. Coolidge invited the choir from the First Congregational Church, which the President and his family attended, to sing carols on the North Portico. The public was allowed to come onto the grounds and hear the carols which Mrs. Coolidge and their son, John, joined in singing.

On January 1st, 1924, the Coolidges resumed the traditional New Year's Day reception that had been interrupted during the war years. She and the President shook hands with three thousand five hundred guests in three and one-half hours.

That summer tragedy struck in the White House again. This time, the Coolidges' younger son, Calvin, sixteen, died from blood poisoning resulting from a blister on his heel which began when he played tennis in shoes without socks. His death put a curb on White House entertaining as well as on his father's campaign for election in November. However, Coolidge was elected without difficulty.

During the next four years, President and Mrs. Coolidge entertained many foreign dignitaries at the White House. Among them was the Prince of Wales (now the Duke of Windsor). The twenty-five-year-old golden-haired prince asked that the luncheon in his honor be informal and it was. The Coolidges chose an all-American menu. It included:

Fruit Cup
Clear Bouillon
Fairy Toast
Speckled Trout
Broiled Chicken
Peas — Bermuda Potatoes
Mixed Salad
Cheese Biscuits
Strawberry Ice Cream
Cakes
Coffee
Salted Almonds — White Rock

Other dignitaries entertained by the Coolidges included Queen Marie of Roumania and her two grown children, Prince Nicholas and Princess Ileana. For them the Coolidges rolled out the proverbial red carpet. A state dinner was held with a reception preceding it in the Blue Room.

The fifty guests were lined up according to protocol in a circle around the room and when the Queen arrived, she went around the circle shaking hands with the Americans. When she came to the wife of a British diplomat who curtseyed, the Queen bowed without shaking hands, and the diplomat bowed and shook her hand. Then the Queen was led to a place at the right of the door opening onto the corridor from which the President and Mrs. Coolidge entered and greeted her.

After dinner, the Queen and Mrs. Coolidge sat on a sofa in the Red Parlor and at each end of the sofa were two gilt chairs where women guests sat, four at a time for a few minutes of talk with the Queen. After the top-ranking women had had their turn, Mrs. Coolidge left and crossed the room to speak to the Queen's lady-in-waiting. In her absence, the Queen asked Alice Longworth to sit beside her and focused her attention on Mrs. Longworth until the men joined them in the Red Room. So it was that some of the other ladies of lesser rank were chagrined that they didn't get to talk to the Queen.

On March 2nd, 1927, the Coolidges moved to the Dupont Circle mansion of Eleanor Patterson, wealthy Washington socialite and newspaper publisher, while a new roof was put on the White House and additional rooms built on the third floor. It was while they lived here that they entertained Colonel Charles A. Lindbergh following his historic flight across the Atlantic.

Mrs. Coolidge attempted to furnish the White House with period pieces and got a resolution through Congress permitting the acceptance of such gifts. However, she had furnished only the Green Room before they left the mansion in March, 1929, when another Republican, Herbert Clark Hoover, became President.

Herbert Hoover

chapter 20

Hoover-Style Hospitality

President Herbert Hoover, a millionaire mining engineer, and his wife, the former Lou Henry, who had met when both were in the same geology class at Stanford University, entertained more often and more lavishly than had any occupants of the White House up to that time.

They rarely sat down to a meal alone. If he invited two guests to dinner, Mrs. Hoover invited a dozen more to make it a party. Not only their official but their private entertaining was marked by good taste, elegance and great expense. Fortunately, whatever his White House allowance did not cover, the President made up from his own private fortune.

Soon after moving into the mansion the Hoovers began to shatter old precedents and set new ones. At their first diplomatic reception they did not have a receiving line. Instead, the Hoovers moved about the room, mingling informally with the guests.

Mrs. Hoover, a linguist who spoke four languages with ease and read two more, stopped and chatted with the diplomats and their wives in their own language as much as possible. She wore a plum-colored velvet dress and brown suede shoes, the same dress she had worn to her husband's inauguration.

It was on Inauguration Day that she planned her first White House party. A large choir had offered to sing for the President and she wanted to invite a cross-section of official and residential Washington to hear them. After shaking hands with fifteen hundred people at a reception following the inauguration, she asked her secretary to make up the guest list and write the invitations by hand so it would seem less formal.

Two months later, Mrs. Hoover began to make plans for entertaining California-style on the White House lawn. A hedged garden one hundred feet square just west of the South Portico was paved with flagstones and provided with white chairs and white tables. This was a favorite spot for entertaining guests at tea during warm weather. Or often, Mrs. Hoover would bring her knitting or reading there in the afternoon, seated in the shade of the spreading magnolia tree.

During their first year in the mansion, the Hoovers gave a dinner in honor of former President and Mrs. Coolidge, and the menu is typical of the table set by the Hoovers. It included:

Consommé
Curled Celery
Rolled Toast Salted Almonds
Brook Trout with Tartar Sauce
White Potato Curls Cucumber Aspic
Chicken Supreme with Spanish Rice
Sweet Potato Croquettes
Grassed Beans Clarified Apples
Hearts of Lettuce and Assorted Cheeses
Toasted Crackers
Vanilla Ice Cream in Molds
Pink Crystal Pears
Homemade Pound Cake
Candies Fruits
Coffee

Here is Mrs. Hoover's recipe for the unusual toast:

ROLLED TOAST

Cut bread very thin so that it will roll without crumbling. Spread with soft butter; roll, starting at one corner, into as tight a roll as possible without breaking the bread. Fasten by piercing with a wooden toothpick; toast slowly; remove toothpick; serve. This may first be dipped into cheese, or spread with cheese, cinnamon and sugar, maple sugar or any conserve.

President and Mrs. Hoover gave their first state dinner on October 7, 1929. It was in honor of British Prime Minister J. Ramsay MacDonald and his daughter, Ishbel. Ninety guests assembled in the Blue Room, decorated with white roses and green ferns, to await the entrance of the President and First Lady. When they arrived they went around the room shaking hands and then led the way to the State Dining Room. Guests were seated at a narrow table in the shape of a horseshoe with the center of the curve banked with palms and ferns.

Another outstanding dinner during the Hoover administration was held in April, 1931, in honor of the King and Queen of Siam. Again the table was in the shape of a horseshoe and lighted by white tapers in gold candelabra, alternating on the table with tall gold stands filled with assorted California fruits and clusters of purple grapes that hung over the sides, almost touching the tablecloth. The Monroe plateau in the center of the table was bedecked with pink tulips, roses, snapdragons and spirea.

After dinner, coffee was served in the Green Room to the ladies, and in the Red Room to the gentlemen.

A suggestion was offered by the State Department for this dinner which called for bowing to the reigning monarch instead of the American custom of shaking hands. However, this grandiose plan was vetoed by the President.

Mrs. Hoover, who seemed to have boundless energy and loved her role, thought nothing of having a luncheon, two teas, and a dinner in the same day. Sometimes she would hold a tea for one group at four o'clock and a second at five, having them in separate rooms so the groups did not merge.

The Hoovers had as their cook in the White House a Marylander named Mary E. Rattley, who had been their cook when the President was Secretary of Commerce and lived on "S" Street in Washington. The President especially liked her original recipes, and she compiled a number of them for publication in a magazine of that day. One was a recipe for cake, which she called Perfect Cake. Here is her recipe:

PERFECT CAKE

1/3 cup butter	2 small cups flour
1 cup sugar	2 teaspoons baking powder
2 eggs	1/2 cup milk

1 teaspoon vanilla

Cream butter, sugar and egg yolks together; sift flour with baking powder; add flour and milk alternately to egg mixture, beating after each addition. Fold in stiffly-beaten egg whites and add vanilla. Turn into well-buttered round mold and bake thirty-five minutes in a slow oven. Fill center with whipped cream and pour hot chocolate sauce over all, or fill center with fruits.

Once her husband's practice of inviting last-minute guests to luncheon or dinner almost created havoc in the kitchen. Four guests were expected for luncheon, and Mrs. Ava Long, the Hoovers' housekeeper, went shopping at eleven o'clock for twelve lamb chops, which she thought sufficient. At twelve-thirty she was notified there would be forty instead of four guests expected at one o'clock!

Mrs. Long knew the unpredictable Hoovers too well to panic but it took some fast action to cope with this situation. She did it by putting the lamb chops through a meat grinder, combining them with ham, beef and almost anything she could find in the larder, and the combination became "croquettes" for forty. They were garnished with mushroom sauce and chopped parsley and served with rice. At that, it was a compliment-winning inspiration. One guest even asked for the recipe, which took some quick thinking by Mrs. Long to concoct again.

Another food crisis was narrowly averted when the Hoovers invited two hundred lawyers to tea one spring afternoon just before they were to leave for their week-end retreat in the Virginia mountains. Picnic baskets of food for their week-end were packed in the trunk of the automobile. But when the guests began to arrive, there were four hundred eighty-five in all. Not enough food had been prepared for the tea tables, so the picnic baskets were returned to the kitchen and the contents converted to tea table fare. Messengers were rushed to the nearest store for more bread, butter, cream cheese and jam, which the kitchen force prepared while butlers were carrying the trays to the tea tables. Consequently, there was enough food for everyone.

Mrs. Herbert Hoover

The Hoover administration was marked by two social storms. One during the first year developed over a social quarrel between Dolly Gann, sister of the Vice President, and Alice Longworth, wife of the Speaker of the House, over rank precedence at White House functions.

Normally, the wife of the Vice President would rank first. But because Mrs. Gann was only a sister of the Vice President, Mrs. Longworth claimed precedence. The Hoovers tactfully settled the matter by adding a dinner for the Vice President to the social calendar, and at this event Mrs. Gann ranked first. Mrs. Longworth then ranked first at the dinner for the Speaker of the House.

Another storm broke when Mrs. Hoover invited the wife of a Negro Congressman, Representative Oscar DePriest, to a tea at the White House for Congressional wives. Many Southern Democratic members of Congress denounced the social mixing of the races in speeches on the floors of Congress, and threatened a social boycott of the White House as long as the Hoovers occupied the mansion. Again the Hoovers quietly rode out the storm without apologies to the dissenters.

After their first year in the White House the Hoovers abandoned one long-standing social tradition, the New Year's Day reception. On January 1, 1930, Mrs. Hoover looked out the window at seven o'clock in the morning and saw the line forming to enter at noon and shake hands with the President.

When the doors were opened nearly four hours later, the President shook hands with three thousand guests before leaving abruptly, while still more guests were climbing the white marble steps to greet him. His hand was bruised and cut where rings had pressed into his flesh.

Whether it was upon the recommendation of the White House physician or whether Mrs. Hoover insisted that he stop is not known, but the following year the White House announced that the President would not be "at home" on New Year's Day, and to assure compliance, the Hoovers left the mansion for the day.

Mrs. Hoover was a doting grandmother who shared with the President his delight in having their two grandchildren, Peggy Ann and Herbert Hoover III, live with them in the White House while their son, who had tuberculosis, was recuperating in a hospital in North Carolina. The First Lady equipped a corner of the upstairs hall with child-size furniture for them, and for any future little guests she might have after her grandchildren returned to their home.

The grandchildren were the center of attention at the Hoovers' first Christmas celebration in the executive mansion. They had a dinner for some fifty guests, and in the center of the horseshoe-shaped table around which they ate, a smaller table was set for the children. On both, there were small brass bells as favors for the ladies, and brass candlesticks for the men. After dinner the President and Peggy Ann, then three, led the guests on a candlelit parade through the darkened rooms, with the girls ringing their bells and the men carrying lighted tapers in their new candlesticks.

These are recipes for foods enjoyed during the holiday season:

PICKLED PEACH ASPIC

(To serve with cold or any wild meat)

1 pint of juice from pickled peaches

1 package of gelatin, well dissolved

Fill chilled aspic molds half full and let stand until firm. Mix cream and Roquefort cheese together, season to taste and place in the half full molds. Add the remainder of the juice.

TOMATO-CHEESE ASPIC

1 quart of juice strained from ripe tomatoes
1 tablespoon sugar
salt to taste
dash of ginger
1 tablespoon vinegar
1 package gelatin, well dissolved

Fill chilled aspic molds half-full and let stand until firm. Mix cream and Roquefort cheese together, season to taste and place in the half full molds. Add the remainder of the juice.

MUSHROOM SOUP

1 pound mushrooms, chopped fine
paprika
1 pint chicken stock
1 pint water, cold
1 pint cream, thin
1 tablespoon flour
1 tablespoons whipped cream

Let mushrooms stand in cold water two hours, then simmer on slow fire about thirty minutes. Strain juice, add chicken stock, well seasoned and add cream and flour mixed in cold milk until smooth. When perfectly smooth, run through a fine strainer. Just before serving, add whipped cream.

That year the Hoovers gave a holiday dance for their younger son, Alan, who was home from Harvard where he was studying business administration. Instead of having the Marine Band, a peppy dance band from a hotel was brought in, the Yuletide decorations were taken down, and in their place fresh spring flowers filled the rooms. About one hundred guests were invited. It was in effect his social debut to Washington, for he had spent the preceding summer at Rapidan Camp in the Virginia mountains.

Mrs. Hoover loved giving unusual parties, and when one of her secretaries got married she gave her a linen shower in the White House. While the guests were having luncheon in the dining room, Mrs. Hoover had a line stretched across the length of the East Room, and, using clothes pins, hung the gifts like wash on a line. After lunch, the guests went into the East Room where the bride-to-be and her friends saw the line full of gifts, and were reminded that Abigail Adams, the first hostess in the White House, had used that very room for hanging her laundry to dry.

Although the Hoovers had lived or traveled in many countries around the world and learned to eat many different types of foods, they preferred to serve American foods in the White House.

Typical foods are:

VENISON

Fashion meat into small cutlets; dip in olive oil and fry in deep fat exactly like doughnuts until done as preferred, perhaps five minutes, but one should experiment with a small bit first. Season with salt and pepper; serve instantly. The cutlet may first be rolled in bread crumbs, corn meal or batter, but his makes them greasier.

MARSHMALLOW SWEET POTATOES

Boil six sweet potatoes and mash through a potato masher to remove strings. Add butter, grated nutmeg, salt to taste, cream enough to make a soft consistency; add one-half pint of ground walnuts. Put in a baking dish, bake for ten minutes. Cover the top with marshmallows and brown as a meringue. Serve at once. Serves six to eight persons.

Present White House Chef Henri Haller mixes ingredients for Eggplant Nicoise, continuing a tradition of good food as served by the Hoovers.

Franklin D. Roosevelt

chapter 21

The Entertaining Roosevelts

During the twelve years Franklin Delano Roosevelt was President, it was nearly "open house" all the time for the friends and acquaintances of the large, gregarious and happy Roosevelt family.

Their entertaining started on Inauguration Day, March 4, 1933, with a buffet luncheon for the Roosevelt family and a few friends as well as a larger reception in the afternoon. But the economic crisis in the nation still in evidence after the stock market crash in 1929, and the sudden death of a senator cause some foreshortening of the ceremonies and celebrations.

However, inaugural festivities reached full peak by 1941, when Franklin Roosevelt was inaugurated for a precedent-shattering third term. There were one thousand two hundred guests for luncheon, served buffet style from two long tables in the East Room and a third table in the State Dining Room. In the afternoon, four thousand additional guests came for tea.

The luncheon consisted of soup, ham, beef, tongue, a salad, cake, ice cream and coffee. Tea guests were also served sandwiches and cake.

Both President Roosevelt, who had never recovered his normal walking abilities since an attack of infantile paralysis at the age of thirty-nine, and his energetic, outgoing wife, Eleanor, then forty-nine, seemed to thrive on association with people. So did their vivacious daughter, Anna, married and the mother of two young children who also lived in the White House, and the four Roosevelt sons, two of whom were still in college when their father became President.

The entire family enjoyed seeing the guests fill the house to overflowing so frequently. Occasionally there were so many, the boys were sent to hotels so their White House rooms could be occupied.

To assist her in dispensing such large-scale hospitality, Eleanor Roosevelt relied heavily on her social secretary, Edith Helm, and her trusted executive housekeeper, Henrietta Nesbitt, who came down from Hyde Park to take on the mammoth job.

Such large-scale entertaining would have been impossible by a less wealthy President since only state dinners and formal parties are paid for out of government funds. Meals for any family or personal friends are paid for by the President personally on a prorata share of the cost.

Mrs. Roosevelt, a shy, sensitive woman who stayed very much in the shadow of her husband until he came into political prominence as Governor of New York, and then President of the United States, was much more interested in social action than in social entertaining, and she was delighted to leave much of the planning to her trusted and capable aides.

When her husband was Governor of New York, he had sent her on inspection visits to state institutions, and she learned to be a keen observer. When he became President, she continued to be the "eyes and ears" for her crippled husband.

She had been in the White House only a few weeks when she took a swing through the country to see what effect the depression was having on individual families. She became the sparkplug as well as the publicity agent for some of her husband's revolutionary programs.

Thanks to a severe economic depression and a difficult Congress, the Hoover years were chaotic and controversial ones. Next to appear on the scene was a man destined to remain in the White House longer than any one before him, FDR.

Eleanor Roosevelt was born into social status and early learned the social graces. So she was right at home in the social atmosphere of the White House. Orphaned at the age of nine, the daughter of Anna Rebecca Hall and Elliot Roosevelt, she and her brother, Hall, were reared by a maternal grandmother, Mrs. Valentine G. Hall, in a palatial ancestral estate on the Hudson River. She was sent to Europe to school, and when she returned to New York, made her debut as was expected of a Roosevelt.

She had many dates but rarely two with the same boy—with the exception of her tall, handsome fifth cousin, Franklin, the only child of James and Sarah Delano Roosevelt, who also lived in a magnificent estate on the Hudson. They were married on March 17, 1905, in a ceremony in which the President of the United States, "Uncle Teddy," Roosevelt, gave away the bride and stole the spotlight from her.

After marriage, Eleanor was dominated by an autocratic mother-in-law, but she meekly found her place in the family and devoted herself to her husband and their five children. When her husband became crippled, and needed her more than ever, she found her opportunities through helping him to achieve a productive and fulfilling life.

Despite her activity in politics and governmental affairs, she never tried to curtail social activity at the White House but found it fitted into the overall pattern of achievement for the Roosevelt administration.

The White House social season during the Roosevelt years started early in December each year and continued until Lent. There was a "little social season" after Lent with garden parties, teas and musicales scheduled into the early summer.

In October of each year, the White House announced dates for the Roosevelts' official entertaining.

It included five state dinners and five official receptions plus additional dinners for chiefs of state whose visits to Washington were sprinkled throughout the season.

At large receptions, the President and Mrs. Roosevelt stood together under the chandelier in the Blue Room to receive, and an aide announced the names of the guests. Color-bearers stood with them, a soldier on one side and a sailor on the other. The guard was changed every twenty minutes.

Cabinet members were not asked to stand with the President and First Lady as it would have slowed the movement of the line of guests, and it was already a strain on the President to stand long enough to shake the hands of all who came. At intervals the receiving was stopped for a few minutes so that the President could sit down and rest. Nearly one thousand six hundred guests were invited to the first diplomatic reception and succeeding ones were almost as large.

No matter how large the reception, the Roosevelts always served refreshments, usually consisting of grape juice, mixed fruit punch, cake, salted nuts and candies.

President Franklin D. Roosevelt welcomes General Charles de Gaulle at the White House on July 6, 1944. Standing behind the President is his daughter, Anna, now Mrs. James Halsted.

President Franklin D. Roosevelt, right, entertains the Crown Prince and Princess of Norway at a tea on the South Balcony of the White House on June 29, 1939.

Favorite Roosevelt dessert recipes were:

BLUEBERRY ICE BOX PUDDING

Line bottom and sides of a glass baking dish with slices of stale bread from which the crusts have been cut. Add a layer of stewed sweetened blueberries. Add alternate layers of bread and berries until the dish is filled, ending with berries on top. Set in the refrigerator overnight. Serve with whipped cream or plain, as desired.

PECAN PIE

3 eggs
1 cup brown sugar
1 tablespoon butter
1 cup Karo syrup
1 cup pecans
1 teaspoon vanilla
Salt, few grains

Cream butter and sugar, add syrup, well beaten eggs, salt and vanilla. When well mixed, add the coarsely chopped pecans, turn into a pie crust. Bake in moderate oven until firm. Walnuts may be substituted for pecans.

President Roosevelt made no effort to stand, however, at the reception he and Mrs. Roosevelt gave for the Washington press corps in the spring of 1934. It was the first time any President had ever honored in this manner the newsmen and women who wrote about the White House and government. It paid off handsomely for the Roosevelts, whose rapport and popularity with reporters has never been equaled before or since.

It was a starlit evening, and about four hundred reporters, writers, editors and their spouses were invited. Upon entering the White House, they were directed to the East Room where the President, wearing a dinner jacket, sat in a comfortable chair at one end of the room. Mrs. Roosevelt stood beside him. Presentations were made by a military aide but from then on everything was informal.

A dance band played in the East Room, and on the East Terrace, through the French doors, were tables and chairs. Mrs. Roosevelt organized the Virginia Reel in the East Room and the President sat and watched the dancing and talked to those who came to him.

In the State Dining Room, ice cream, cake and coffee were served. In the main hall were kegs of beer. It was a memorable and historic night for the press corps and one that was repeated in slightly varied forms in subsequent years of the Roosevelt administration, before the country was plunged into war and entertaining was once again curtailed.

The following December, Mrs. Roosevelt began a tradition for the women of the press and wives of newsmen. Every year the Gridiron Club, an organization of newspapermen, gave a dinner at a Washington hotel for the President, followed by a program of original skits written by the members, spoofing public figures. It was a "stag" dinner and no women were allowed. So Mrs. Roosevelt retaliated by organizing the "Gridiron Widows' Party," which she held at the White House on the night of the Gridiron dinner.

The first "Gridiron Widows' Party" was a masquerade, with prizes given for the best costumes. As part of the entertainment, Mrs. Roosevelt and her daughter, Anna, presented an original skit. Mrs. Roosevelt took the role of an elderly lady selling apples. Dressed in a shawl and sunbonnet, she was mumbling about her woes, when Anna, dressed in a red devil suit, appeared and transformed her into a First Lady for a day. After weighing the problems involved in each, Mrs. Roosevelt decided, in the skit, that her life was easier selling apples and gladly reverted to that role.

It drew many laughs and hearty applause from the women who recognized more than anyone else, perhaps, the multiple problems faced daily by the First Lady, and the criticism she encountered from many sides for her pioneering role in trying to help the nation's downtrodden.

At the Roosevelt's state dinner for the Cabinet that year, wine glasses made a reappearance on the table in the State Dining Room for the first time since the repeal of prohibition. Light wines of American vintage had been served on occasion prior to repeal, however.

For this cabinet dinner, the table was decorated with the Monroe gold service banked with pink roses, pink snapdragons, and stevia with feathery maidenhair fern. Spaced at intervals on the horseshoe-shaped table, overlaid with fine white damask, were twelve bowls of pink and white flowers.

After the Roosevelts' state dinners, there was always music in the East Room, with additional guests, sometimes as many as two hundred, invited to come at ten o'clock for the musical program. Mrs. Roosevelt greeted them in the Green Room before leading them to the East Room, where they joined the dinner guests on small gilt chairs. Usually after the concert, cookies and punch were served in the State Dining Room, and occasionally, a late supper for the performing artists and guests.

One concert ended with a midnight supper in honor of the artist. On this occasion the menu included:

Chicken Consommè
Filet of Lamb
Green Peas   Romaine and Grapefruit Salad
Orange Blossom Parfait
Angel Cake
Coffee

Among the first dinners given by the Roosevelts that first spring in the White House, was one in honor of Paderewski, the Premier of Poland. On the menu for the twenty guests were:

Fruit Cocktail
Cream of Celery Soup
Planked Shad Roe   Swiss Dressing   Sliced Cucumber
Crown Roast   Mint Jelly
Green Peas   Scalloped Corn   Pan-Roast Potatoes
Lettuce Hearts   Cream Cheese   Bar-le-duc
Strawberry Cream
Cake   Coffee

In planning state dinners, religious and dietary practices of the guests of honor were taken into consideration. Ras Desta Demtu, the Abyssinian ambassador, who was a guest at the White House in the summer of 1933, was a Coptic Christian and did not eat meat, butter, cheese or any product of animal food. The menu for him got around the restrictions by including:

Melon
Clam Cocktail
Saltines — Stuffed Olives — Ripe Olives
Prime Filet of Bluefish
Grilled Tomatoes — Mexican Corn — Molded Potatoes
Cucumber and Cress Salad — Beaten Biscuit
Pineapple Ice
Stuffed Dates — Ripe Figs
Coffee

Christmas meant a traditional family celebration with the Roosevelts. The President always read Charles Dickens' "Christmas Carol" to his children, even after they were grown up, putting his own dramatic emphasis on the role of Scrooge. All year long Mrs. Roosevelt shopped for the very personal, useful gifts she wrapped for her family and close friends. Usually there was a room full of children, grandchildren and guests when they finally sat down late in the day for the holiday meal.

President Franklin D. Roosevelt, seated second from right, Mrs. Roosevelt, left, and his mother, second from left, with some of their children and grandchildren before the Christmas Tree in the East Room on December 25, 1939.

In 1933, the Christmas dinner menu was an elaborate one. It included:

Clam Cocktail Saltines
Clear Soup
Beaten Biscuit
Curled Celery Stuffed Olives
Filet of Fish
Sauce Marèchale
Sliced Cucumbers
Rolls
Roast Turkey
Chestnut Dressing
Deerfoot Sausage
Cranberry Jelly
Creamed Onions Green Beans
Candied Sweet Potatoes
Cheese Straws
Plum Pudding
Hard Sauce
Ice Cream
Small Cakes Cookies
Coffee
Candy

In later years, and especially during World War II, the menu, while still festive, was streamlined considerably.

Just before Christmas in 1934, the President and Mrs. Roosevelt welcomed their two sons, Franklin, Jr., and John, home from school for the holidays with a dinner dance. The mansion was decked in holiday dress. Four Christmas trees, brilliantly lighted and shining under the white portico facing Pennsylvania Avenue, welcomed the three hundred guests as they arrived.

In the East Room where the dance was held, a Christmas tree stood in the recessed windows, and Christmas greens and holly, interspersed with red poinsettias, banked the four mantels.

Sharing honors that evening were two young friends of the Roosevelt boys, who were guests at the White House. They were Miss Jean Martineau, from England, who was spending the winter in New York with the President's sister, Mrs. Price Collier, and Barbara Cushing, sister of James Roosevelt's wife.

President and Mrs. Roosevelt, their two sons and the two house guests received the party guests in the Blue Room. The President remained for some time afterwards, talking with various people and watching the dancing. Later in the evening, Eddie Peabody, a popular young banjo player, gave a performance and played also during the supper, which was served from a buffet in the State Dining Room.

Franklin Jr. and John made out a list of popular songs which they asked Meyer Davis and his orchestra to play at the dance. The list began with one of their father's favorites, "Home on the Range," and continued with "The Continental," "Harvard Glide" (a fox trot written by Davis in honor of the young hosts), "You're the Top," "Merry Widow" waltzes, "Anything Goes," "Two Cigarettes in the Dark," "Stars Fell on Alabama," "Flirtation Waltz," "June in January," "An Earful of Music," "Stay as Sweet as You Are," "The Blue Danube," and "The End of a Perfect Day."

Mrs. Roosevelt gave a number of other White House dances, some for her own sons, and others for relatives and close friends. She gave a debut dance for her young niece and namesake, Eleanor Roosevelt, on Saturday, December 30, 1933, to which seven hundred were invited. Each year she gave a children's party for her two grandchildren, Sistie (Anna Eleanor) and Buzzie (Curtis) Dall, Anna's children, who lived in the White House. And other grandchildren were invited as they arrived through the years. The entertainment varied each year. Sometimes it was a Punch and Judy show, sometimes a dance group, or a famous personality from the entertainment world. It was not difficult to get a top name personality from show business to give a performance at the White House, regardless of the occasion.

Sunday nights were special events for the Roosevelts, because they often meant an informal dinner party. The score or more guests would include half a dozen newsmen and their wives, perhaps an ambassador or two, and half a dozen government officials from the second and third echelon, and their wives.

President Roosevelt sat at one end of the table, and Mrs. Roosevelt seated at the other, scrambled eggs in a chafing dish to complete the menu of bacon or sausage, salad, coffee and dessert. After dinner, the guests sat clustered around the President, and it was not unusual for some to sit on the floor to get closer to the center of the conversation. There was a lively exchange of ideas, comments and jokes. It amounted to a scrambled brains session as well as a scrambled eggs supper.

Here are some of Mrs. Roosevelt's simple supper recipes, which were served in the White House and at the family home in Hyde Park:

TOASTED CHEESE SANDWICH

1 pound of sharp cheese grated — white sauce

Make thin white sauce, add grated cheese gradually and cook until thoroughly melted to a smooth consistency. Sprinkle with paprika and serve on slices of toast.

KEDGEREE

1 cup white fish, boiled and flaked
1 cup rice, boiled
2 eggs, hard boiled, minced
Salt and pepper to taste

Mix fish with other ingredients, add milk to make the mixture slightly moist. Put in the oven and brown. Serve hot.

A dancing class was organized at the White House by Mrs. Roosevelt and met twice a month during the early years in the East Room for square and round dancing. This group included some of the same guests who were invited to the Sunday night suppers.

Mrs. Roosevelt gave parties in a series, sometimes three teas or receptions in one afternoon, with several hundred guests at each. Tea was served from a lace-covered table centered with a bowl of flowers, and the menu consisted of cinnamon toast, tiny sandwiches, cookies, little cakes, candies and nuts.

With such frequent and large-scale entertaining there were occasional mishaps. Once, three hundred were expected for tea, the table was set, all arrangements made but not a single person came. Someone had forgotten to mail the invitations! The food was

Typical of the floral arrangements used to decorate the White House in the 1960's is this bright spray of white shasta daisies, pink mums and wine-colored asters on a velvet-covered table in the Blue Room beneath the handsome crystal chandelier.

An artistic arrangement of colorful black-eyed Susans, daisies, snapdragons, lilies and iris decorates the center of the round table in the oval Blue Room where receptions are held.

sent to a local charitable institution. Another time, for a diplomatic dinner, Mrs. Roosevelt had an artist design individual place cards for the guests with a scene symbolic of their home-country. For one guest, the artist drew large birds eating in a city street. Luckily, a State Department aide saw it in time and removed it, for the birds to the diplomat meant buzzards, a true reminder of his city but not a pleasant one. After that Mrs. Roosevelt reverted to the gold seal on the place cards.

President Roosevelt's international relations almost got off to a bad start, all because of the family's pet, a German shepherd named "Major." John Roosevelt, the President's youngest son, tells the story today. The dog had been trained by the New York State Police, when the President was Governor and living in the executive mansion in Albany. One evening the President and First Lady gave a dinner at the White House in honor of three visiting foreign ministers of different countries. As the President and Mrs. Roosevelt received the three honor guests, "Major" sat nearby at attention. When the third minister, a continental gentleman, bowed from the waist to kiss the First Lady's hand, "Major" made one leap and tore the seat from the foreign minister's pants. For a split second everyone stood electrified, then, while one person grabbed the dog, several others quickly ushered the "attacked" minister out of thr room and to the butler's pantry, where the butlers were lined up in evening dress to begin serving dinner. One was spotted in the line approximately the size of the minister whose trousers had been deseated. The butler removed his trousers and the minister put them on, returning to the scene of the debacle. Apologies were profuse, and soon "Major" was returned to the New York State Police.

The most spectacular dinner given by the Franklin Delano Roosevelts during their twelve years in the mansion was the state dinner in honor of England's King George VI and Queen Elizabeth, the first British monarchs to set foot on American soil.

It was on June 8, 1939, a typically-hot Washington summer day. The royal couple arrived by train just before luncheon and rode motorcade-fashion in a slow moving open car from Union Station to the White

A musical group from the Marine Band plays for an afternoon party at the White House in the 1960s.

House. By pre-arrangement both women (the Queen and Mrs. Roosevelt) were to wear wool dresses, despite the summer weather, to give British wool a boost. As it turned out, Mrs. Roosevelt wore her wool and was very uncomfortable in the sweltering weather. However, the Queen had a change of heart and wore something cooler. She held a ruffled parasol over her head during the motorcade to keep the sun off her smooth, white skin, but even so, the noonday sun brought a blush of pink to her face before she reached the White House.

There, a private luncheon was served the royal couple. It included minted melon balls, green turtle soup, broiled sweet breads and mushrooms, asparagus, saratoga chips, hearts of lettuce, salad with roquefort dressing, pineapple shortcake, coffee, nuts and candies.

Weeks of preparations and trans-Atlantic communications preceded the state dinner given that evening in honor of the royal pair. The President and Mrs. Roosevelt wanted to give them the best in American food and a cross-section of the best in American entertainment, compatible with their personal preferences.

That evening, the great horseshoe table in the State Dining Room had never been more beautiful. The Monroe mirror plateau was used, and the great épergnes were edged with clusters of golden grapes. Purple-tipped white orchids were massed in gold bowls, white candles gleamed in gold candelabra, and bouquets of roses, baby's breath and fern were attractively spaced along the white damask-covered table. The portrait of Abraham Lincoln, hanging above the mantel, was decorated with white gladioli, and low green palms were placed in the center of the horseshoe. Flowers bloomed throughout the mansion. Tall vases of purple gladioli were even placed beside each column on the South portico, where the guests went during the evening for a breath of cool air.

The White House china selected by President Roosevelt was used. It was Lennox white with a cobalt blue border and feather design in gold, marked with the presidential coat of arms and the Roosevelt crest.

Across the beautiful table President Roosevelt and a smiling King George toasted each other, and eighty guests raised their sparkling champagne glasses and drank. In toasting the tanned young monarch, President Roosevelt cited as an example to the world the "methods of peace, divorced from aggression," which the United States and the British empire have followed in their relations with each other through their many years of friendship.

King George responded in clear tones with no trace of the stutter that sometimes beset his speech. "I hope our great nations may ever in the future walk along the path of friendship in a world of peace," he said. He wore his much-decorated uniform and white gloves. President Roosevelt looked just as impressive in his white tie and formal jacket with tails.

Queen Elizabeth wore a Norman Hartnell dress of sheer white tulle, the bodice formed of white, embroidered with gold pailettes. The full skirt was three tiers of white tulle powdered with gold-sequined embroidery in graduated design. In her soft dark hair sparkled a small tiara set with pearl-shaped rubies encircled by diamonds. Her necklace, earrings, and bracelets were of matching design.

Mrs. Roosevelt wore a gown of natural-colored Alencon lace designed especially for the occasion. It was closely fitted to the hipline, then flared into a full skirt. Jeweled clips drew the neckline into a heart shape, and lace flowing from the shoulders formed her cape sleeves. Her jewels were a string of graduated diamonds which had been in the Roosevelt family for generations, and a large emerald-cut diamond engagement ring.

President Roosevelt sat on the outside curve of the horseshoe with Queen Elizabeth at his right, and Mrs. John Nance Garner, wife of the Vice President, at his left. Directly across the table from him sat Mrs. Roosevelt with the King on her right and the Vice President on her left.

Guests included cabinet members, high army and navy officers, select members of Congress and the State Department, and the wives of two men who had battled for the presidency: Mrs. Woodrow Wilson, whose husband won the prize; and Mrs. Charles Evans Hughes, whose husband became the Chief Justice. Her husband was ill, and unable to attend the dinner. Also among the guests were the President's secretary, Marguerite LeHand, G. Hall Roosevelt, the First Lady's brother,

Malvina Thompson, the First Lady's secretary, Mrs. James Meredith Helm, the social secretary, James, Elliot and Franklin D. Roosevelt Jr., and their wives.

The dinner menu selected for the King and Queen included:

Clam Cocktail
Calf's Head Soup
Terrapin Corn Bread
Boned Capon
Cranberry Sauce
Peas Buttered Beets
Sweet Potato Cones
Frozen Cheese and Cress Salad
Maple and Almond Ice Cream
White Pound Cake
Coffee
Wine

The king sampled everything. However, the Queen, who was suffering from the heat, tasted only the terrapin and capon, beets, and drank her coffee.

Just as the guests sat down at the table, lightening ripped the summer sky, and a violent downpour lashed the exterior of the White House. It was an unexpected display of nature, but a welcome one. It did little to cool the interior of the mansion however, which was not yet air-conditioned.

After dinner, the guests were joined by two hundred more, invited for the musical program in the East Room. At one point, the heat became so oppressive in the room that the Queen was led outside on the South Portico to be refreshed by the cooling breezes brought by the rain.

The entertainment ranged from opera singers to hillbilly dancers and folksingers. At the special request of the King, Kate Smith sang her radio theme song, "When the Moon Comes Over the Mountain," and two other selections: "Macusha," and "These Foolish Things." Marian Anderson, the famed contralto, sang "Ava Maria," "My Soul's Been Anchored in The Lord," and "Tramping." Lawrence Tibbett, baritone, sang "The Pilgrim's Son," "Sylvia," and "If Love Hath Entered My Heart." Alan Lomax, of Washington, D. C., compiler of thousands of folk songs, presented "Whoopee, Ti, Yi, Yo," "Git Along, Little Doggie," and "The Old Chisholm Trail."

From North Carolina came The Spiritual Singers, who sang "De Ol' Ark's A Moverin!", "Wade In the Water," and "I Don't Feel No Ways Tired."

Four young women from Coon Creek, Kentucky, calling themselves the Coon Creek Four, and including Rose Ledford, twenty-three, Lily May Ledford, twenty-two, Violet Koehler, twenty-three, and Daisy Lang, nineteen, played musical instruments and sang "How Many Biscuits Can You Eat," and "Cindy." One of the four, Lily May, told the audience she kept a snake's rattlers in her violin to improve the tone.

Examples of authentic mountaineer dancing were given by the Soco Gap Square Dancers from Asheville, North Carolina. The group that included farmers, mechanics, school teachers and housewives, dug hard-bottomed shoes into the East Room platform with wild careening spasms to renditions of "Dive and Shoot the Owl" and "London Bridge."

It may have been a new form of entertaining for the King and Queen of England, but President and Mrs. Roosevelt's dinner guests had been initiated into such musical drama before. One of their earlier after-dinner programs had featured a group from White Top Mountain in Virginia. Dressed in their Sunday best (neat black suits with white shirts and black ties for the men, black dresses with white collars for the women), the group provoked interest not only with their performance but their ages. Miss Betty Smith, age eighty, fiddled "Hell Broke Loose in Georgia" as a solo, and "Haste to the Wedding" in duet with her brother, Jim Chisholm, seventy-six. The songs that appealed most to the President on this occasion, however, were the sea chanties including "Santy Anno" sung in rollicking style by grizzled, gray, banjo-playing J. M. Hunt who had sailed in clipper ships in his youth.

During World War II, large-scale entertaining ceased and there was no official social season. However, the parade of important visitors to the White House continued and there were dinners, luncheons and teas for them. Prime Minister Winston Churchill was a visitor many times, once on Christmas Day. Other important visitors who received red-carpet welcomes included King George of Greece, King Peter II of Yugoslavia, Queen Wilhelmina of the Netherlands, Crown Princess Martha of Norway, her husband, and many others.

While her husband carried on the affairs of state, served as commander in chief of the nation's military forces, and Allied leader, Mrs. Roosevelt sandwiched her White House hostess duties in, between travels to battle zones around the world as "eyes and ears" for the President.

Although the White House was closed to tourists, she gave tours for the returning wounded servicemen at Washington area hospitals. She personally showed them the Lincoln portrait painted by Healy in Rome in 1869, hanging over the mantel in the State Dining Room. She told them that it was acquired by the President's son, Robert Todd Lincoln, who willed it to the White House. It did not arrive until the Roosevelt administration, however, because Mrs. Robert Todd Lincoln had it during her lifetime.

Mrs. Roosevelt also pointed out the prayer John Adams had written to his wife Abigail on his second night in the White House in 1800. President Roosevelt had the prayer engraved in the marble above the fireplace in the State Dining Room as a reminder to all who should occupy the mansion thereafter. It reads:

I Pray Heaven to Bestow
The Best of Blessings on
THIS HOUSE
and on All that shall hereafter
Inhabit it. May none but Honest
and Wise Men ever rule under This Roof.

The Roosevelt rule, during years of war and progress, ended on April 12, 1945, with the sudden death of the President at Warm Springs, Georgia, five months following his election to an unprecedented fourth term.

Egg rolling on the White House lawn on Easter Monday is a custom brought to the mansion by President Rutherford B. Hayes and continues today.

White House floral designer, Elmer (Rusty) Young arranges red, white and blue flowers in vermiel bamboo containers for tables at a state dinner in the summer of 1967. Through the years each administration has selected the floral decorations they wanted for the mansion.

Harry S. Truman

chapter 22

The Lights Go On Again

It was many months after Vice President Harry S. Truman became President in April, 1945, succeeding the late Franklin D. Roosevelt, that the lights for official entertaining came on in the White House again.

For nearly five long years the mansion had been dark to official entertaining, because of the war which did not end until the August after Truman became President. Even then, many American servicemen were still abroad and many American families still mourned the loss of loved ones.

Consequently, President Truman, the hardworking former haberdasher from Independence, Missouri, did not feel it was proper to open the White House for official entertaining.

It was November of the following year, 1946, that the first postwar social season opened in the White

House. Thereafter, nearly every Tuesday night that winter, windows in the White House blazed with light indicating there were festivities inside.

Two diplomatic dinners were held that year, because during the war years and immediately afterwards, the diplomatic corps had grown too large for one gathering. There were now sixty chiefs of mission, compared to fifty-three in 1939, and if each brought his wife, the number would exceed the capacity of the State Dining Room.

As a result, the diplomats listed in the State Department's blue book were numbered according to length of service in Washington. The even numbers were invited to the Trumans' first dinner for the diplomatic corps, and the odd numbers to the second, the following Tuesday.

Knowing that ambassadors are very jealous of rank and position, the White House let it be known how the guests were selected for each dinner, and even went so far as to measure by inches the distance the ambassador of each country sat from the President.

For the first dinner, the table was decorated with pink carnations and pink and white pompom chrysanthemums mingled with Baker's fern in tall golden vases set on the Monroe mirror plateau. Among the guests were General of the Army Dwight D. Eisenhower and Mrs. Eisenhower, who seven years later, were entertaining in the White House themselves as President and First Lady.

Following the dinner, a musical program was presented in the East Room by Sylvia Zaremba, a fifteen-year-old pianist from Chicopee, Massachusetts.

The second diplomatic dinner was a carbon copy of the first, except that the entertainment was provided by Lawrence Tibbett, the Metropolitan Opera baritone.

Four more dinners and five receptions were given that season by President Truman and his gentle, friendly wife, Bess Wallace Truman, who had been his childhood sweetheart. All the dinners and receptions followed the same pattern with slight variations in the entertainment.

At the dinner for the cabinet on December 17, 1946, the first one attended by Margaret, the Trumans' twenty-two-year-old daughter who had graduated from George Washington University, the entertainment included songs by Helen Traubel, the Metropolitan Opera's dramatic soprano. Following the dinner for the judiciary a few weeks later, the guests heard pianist Oscar Levant.

The receptions likewise followed a pre-set pattern. About seven hundred guests lined up four abreast in the East Room and then merged into a single line before entering the Blue Room to be received by the President and Mrs. Truman. The Trumans had emerged from the upstairs family apartment promptly at nine as the Marine Band played "Hail to the Chief."

They were followed by members of the Cabinet who stood in the Red Room with members of the President's official family, or in the Green Room where specially invited guests assembled earlier. It was considered a distinction to be among those guests in the Green and Red Rooms at receptions.

Fifteen minutes before the President and First Lady came down the stairs, four young servicemen—an airman, a Marine, a soldier and a sailor—were taken by a military aide up to the President's oval study where the President sat at his desk, flanked by two flags, the "stars and stripes" and the President's own flag.

Then followed a little-known but impressive ceremony witnessed by few outside the Presidential family. The aide salutes the President and asks permission to remove the flags. The four servicemen stand at attention, then two of them remove the colors from standards and march back down the stairs where they take positions, one at the right and other at the left of the door to the Blue Room. There they stand at attention as the President and First Lady walk through, remaining frozen for a twenty minute interval, after which they are relieved by the two other military guards. The four alternate in that spot until the end of the reception when they take the colors back upstairs.

At the January 7, 1947 reception for the diplomatic corps, the chief topic of conversation was the resignation, announced just a few hours earlier by the

President, of James F. Byrnes of South Carolina as Secretary of State, and the appointment of General George C. Marshall, wartime Chief of Staff, to succeed him.

"Is there anything political about this?" the diplomats and government officials asked each other. The outgoing Secretary of State and Mrs. Byrnes held court, as ambassadors and Senators filed by to express their regrets. The Secretary was late arriving, because he was detained at home receiving telephone calls concerning his resignation.

Refreshments were always served at the Trumans' receptions, and included punch, cake, cookies, chocolates and nuts served from a buffet table in the State Dining Room. Bowls of pink roses and baby's breath decorated the table, and flame-colored gladioli stemmed from every corner.

Although President and Mrs. Truman decided not to have any formal receptions or dinners during the winter of 1945, Mrs. Truman continued the ladies luncheons that had been started by Mrs. Roosevelt. That autumn also, she joined with wives of senators, (a group familiar to her through long association,) in giving a luncheon at the White House for one hundred servicemen in area hospitals.

Mrs. Truman, who had worked in her husband's office when he was a Senator, took a business-like approach to running the White House. She disliked the spotlight beamed on the First Lady and preferred to remain in the background giving support and counsel to her peppery, outspoken husband, who referred to her as "the boss." It was during the Truman administration that the president's salary was increased from $75,000 to $100,000 a year.

Mrs. Truman enjoyed small, informal gatherings with the women she knew well more than the formal dinners and receptions at the White House, although she performed her hostess duties at the latter with grace and dignity.

In the spring of 1946, she invited her bridge club from Independence, Missouri, to the White House for a visit, and literally rolled out the red carpet for the delighted middle-aged women. They went sightseeing together, took swims in the White House pool (built early in the Roosevelt administration with contributions from children all over the country), and were guests of honor at a luncheon by Mrs. Truman.

Another group Mrs. Truman enjoyed especially was her Spanish class, taught by Professor Ramon Ramos which met in the White House. Once each year it came her turn to have the monthly luncheon for the Professor's five classes, including her own. Her first luncheon for the group was held on April 20, 1946, and it was a memorable day in the White House kitchen. The professor and nearly a dozen of his students took over at nine in the morning to prepare the luncheon, which had a Latin American flavor in honor of Pan American Day.

The chief dish, which took three hours to prepare, was called Picadillo (see recipe below) and was a special concoction created by Professor Ramos. Mrs. Truman helped cut up the beef, pork and lamb that went into the dish. Then, under his tutelage in Spanish, mixed the meat and rice, seasoned it with hot spices and garlic, and garnished it with almonds, pimentos, olives and raisins.

She was assisted by the wives of some of the capital's most prominent men. They included Mrs. Dean Acheson, wife of the Under Secretary of State; Mrs. Leverett Saltonstall, wife of the Massachusetts Senator; Mrs. Lester Pearson, wife of the then Canadian Ambassador, Representative Jessie Sumner, a Republican from Illinois, and Reathal Odum, Mrs. Truman's secretary.

During the preparation the cooks sang Spanish songs and learned the Spanish words for the culinary terms and equipment they used.

A little later the waitresses came in, and a famous company they were: Mrs. Dwight D. Eisenhower, Mrs. Hugo Black, wife of the Supreme Court Justice, Mrs. Robert Patterson, wife of the Secretary of War; Mrs. George Allen, wife of the head of the Reconstruction Finance Corporation; Mrs. John L. Sullivan, wife of the Assistant Secretary of the Navy; and Mrs. Brien McMahon, wife of the Democratic Senator from Connecticut.

Wearing fancy little aprons, they served sixty-six classmates seated at the horseshoe table in the State Dining Room.

PICADILLO A LA RAMOS

1 pound lean round steak
½ pound lean pork
½ pound veal
2 cups dry white wine
½ cup raw, shelled almonds
1 cup pitted olives
1 cup raisins
2 large tomatoes
2 hardboiled eggs
2 cloves garlic
1 large green pepper
1 medium-sized onion saffron
several threads of toasted saffron
several sprigs of parsley
pinch of cumin seed
2 bay leaves
salt to taste

Cut the three meats into small pieces and grind together to mix well. Chop the tomatoes, garlic, onion and pepper and brown in a frying pan with a little butter. Add the ground meat to the pan and continue frying for ten minutes, stirring constantly. Put the contents of the pan into a pot with the almonds, olives, bay leaves, cumin seed, wine, and sufficient saffron to color the dish. Simmer for one hour, adding water as is necessary to keep the dish juicy. Add the raisins fifteen minutes before the end of the hour. Serve over white rice. Decorate each serving with two slices of hard-boiled egg and a sprig of parsley. Serves six.

In addition to Picadillo, the Cuban meat dish mentioned above, the menu included a green salad and a dessert of New Mexican cheese served with guavas in syrup.

After luncheon, the ladies gathered in the East Room and sang songs in Spanish.

It was an unusual White House party and one the guests thoroughly enjoyed. Although the class members did the cooking and serving, they were relieved of the chore of cleaning up. The White House kitchen staff, which had gotten out the pots and pans early in the morning, took over and put things in order again.

The affair was so successful it was repeated during Pan American Week the following year.

After one successful social season, the Trumans curtailed formal entertaining in 1948 because of hunger abroad. While the United States was sending shipload after shipload of wheat and footstuffs abroad to aid war-devastated nations, it did not seem right to feast in the White House, so all state dinners were eliminated.

The Trumans continued official receptions, however, as they did not require great outlays of food. It was at one of these that President Truman noticed an unusual tinkling of the crystal teardrops of the chandelier in the Blue Room and asked for an investigation to determine the cause. It turned out that the White House was a structural hazard after one hundred and forty-eight years, and Congress soon authorized a five million dollar rebuilding program.

After the 1948 election in which President Truman won another four years in office, the First Family moved across the street to Blair House, used as a Presidential guest house, and lived there during the three years the Executive Mansion was being rebuilt.

The Trumans entertained on a limited scale at Blair House, but large official functions were held at various local hotels. They gave a state dinner in honor of the President of Chile, Gabriel Gonzales Videla, his wife and their daughter at the Carlton Hotel. But when Princess Elizabeth, the fifth member of British royalty to visit the United States, and her husband, Prince Philip, came on October 31, 1951, the Trumans gave a small dinner party for them at Blair House. There were only eighteen guests that evening, for whom a segment of the Marine Band played music. Later, an additional hundred guests were invited to a reception to meet the royal visitors.

Mrs. Truman continued her buffet luncheons for ladies at Blair House. Often these were held in the little patio garden at the rear, and the menu usually ran something like this:

Jellied Madrilene
Cold Sliced Tongue
Chicken and Ham
Fish Salad
Mixed Vegetable Salad   Potato Chips
Celery   Carrot Fingers   Olives
Corn Sticks
Ice Cream   Cake
Coffee

Following the outbreak of the Korean War in 1950, Mrs. Truman started having afternoon teas at Blair House for hospitalized servicemen. She invited some of Margaret's friends to assist her as hostess. The First Lady received the men in the drawing room and refreshments were served in the dining room while a Marine "combo" played in the hall. President Truman almost always dropped in at these parties.

Here are recipes for some of the goodies Mrs. Truman served in the White House:

LEMON PIE

2 eggs, separated
1/3 cup fresh lemon juice
1 tablespoon grated lemon rind
1 cup whipped cream
1/2 cup sugar
1/2 cup crumbled graham crackers

Beat egg yolks. Add lemon juice, rind and all, with two tablespoons of the sugar. Cook over low heat, stirring constantly. Cool. Beat egg whites, add two more tablespoons of sugar, fold into cooked mixture; then fold in whipped cream. Line greased pie or refrigerator pan with graham cracker crumbs. Save some to sprinkle on top. Pour filling into pan and freeze. Serves four.

CHEESE RING

2 small packages cream cheese
1 cup grated yellow American cheese
dash of salt
2 tablespoons gelatin dissolved in cold water
1 cup whipped cream
1 cup milk

Add dissolved gelatin to heated milk. Mix and cool. Mash cream cheese with fork and add yellow cheese. Combine with gelatin mixture. Let thicken and add cream. Use as a base for congealed fruit salad or place in mold until set. Turn out on plate and surround with fresh fruit.

Margaret Truman, a pretty, blue-eyed blond who had not yet developed the glamor she has today, was her parent's pet and pride. They were a close-knit family who shared each other's joys and problems. She found life in the White House somewhat restrictive when it came to romance, although she never lacked for dates and escorts.

She entertained often at small parties for her close friends, the entertaining frequently including the viewing of a new or favorite old movie in the White House Theater. When Gloria Chavez, a Senate daughter with whom she had grown up, became engaged to be married, Margaret gave a party for her in the White House. She gave several dances also. However, like her parents, Margaret was not the flamboyant or dramatic type, and her parties were planned more for fun and fellowship than originality and showmanship.

After graduating from college, Margaret went to New York to study voice, and then began her own career as a concert singer which took her to cities throughout the United States. But she always kept in close touch with her two favorite critics and boosters, the President and Mrs. Truman.

In 1952, nine months before he left office, President Truman and his family moved back into the rebuilt White House. It was much larger, now having one hundred and thirty-two rooms, with many new conveniences such as air conditioning, extra bathrooms, adequate clothes closets and even an air-raid shelter in the basement.

When the White House was reopened to formal entertaining, the Trumans decided to hold their receptions in the late afternoon, starting at five o'clock instead of nine in the evening. A number of afternoon receptions were held in April and May, 1952, and so many people came to see the new White House, the Trumans topped all previous crowd statistics for a single season of entertaining.

The guests saw the beautiful new East Room, which was panelled in wood to the ceiling, and the windows, where cranberry-red drapes had formerly hung, were now covered with white and lemon-gold silk damask drapes of authentic eighteenth century design.

The Green Room had been completely redecorated, with a love seat and two occasional chairs upholstered in a much lighter green than the walls; the Blue Room walls were covered with a new blue silk satin with a large gold decorative motif, and the furniture was now white, blue and gold. The Red Room was toned down by having three chairs upholstered in off-white, where once all had been red.

When President Truman was succeeded on January 20, 1953, by President Dwight D. Eisenhower, Mrs. Truman had no regrets about moving out of the White House spotlight and back to her comfortable, quiet home in Independence where she resumed the role she cherished most, wife and mother. Not many years later in the little Episcopal Church in Independence she saw her only daughter, Margaret, marry the man who won her heart, E. Clifton Daniel. And a few years later President Truman achieved the role he seemed to cherish more than being President. He became a grandfather.

The State Dining Room with tables set for 140, the maximum number of guests possible in this room for a state dinner.

Dwight D. Eisenhower

chapter 23

The Eisenhower Era

When Dwight D. Eisenhower, a retired Army General and hero of World War II, became President of the United States and moved into the executive mansion with his vivacious, outgoing wife, Mamie, it marked the beginning of a new era of entertaining in the White House.

No previous First Lady enjoyed living in the White House more than Mamie Eisenhower, who had made her home in scores of apartments, flats, houses and mansions during her thirty-seven years as an Army wife. She had been a guest at the White House many times and was accustomed to official entertaining. She loved people, and throughout her married life to the popular Army officer, their home, wherever it was, had been open to their many friends, old and new.

From the end of World War II, the Eisenhowers had moved in top official circles in this country and abroad, and while he was Supreme Commander of the North Atlantic Treaty Organization, she traveled extensively abroad with him and had been entertained by virtually every royal family or head of government. So the official hostess role she acquired when General Eisenhower became President came easy to her. So did meeting people from other countries and all walks of life.

Mrs. Dwight D. Eisenhower

As First Lady, Mamie Eisenhower was eager to make new friends and she never forgot her old ones. She continued to play cards (usually canasta or bolivia) at the White House with her friends from Army days. Most of them were wives of prominent Army officers or public officials. They would sit down at the card table at two o'clock in the afternoon and play until five. They stopped for tea and then began again and played until dinner. Occasionally they would take a break for a movie after tea, and if it was a cowboy-and-Indian film, the President joined them when his duties permitted.

Even at sixty Mamie Eisenhower had clear, unlined skin, pink cheeks, blue eyes and a size fourteen figure that was suited to the feminine fashions she loved. She had never enjoyed robust health, and she guarded it closely in the White House, resting in bed until late every morning to conserve her energy for the afternoon and evening social events.

Because Mrs. Eisenhower enjoyed living in the White House (and she lived there longer than any place during her first thirty-seven years of marriage), she tried to make her guests enjoy it as much as she did. Her first concern, however, was her husband, and she looked after his comfort first.

During their eight years in the White House the Eisenhowers entertained more foreign chiefs of state than had any other president, and they established precedents in giving other parties also.

For example, in May, 1959, Mrs. Eisenhower gave a garden party for six hundred of Washington's senior citizens who lived in homes for the aged. They came from twenty-four institutions, some in wheel chairs, some on canes, and some with a spring in their step. Mrs. Eisenhower greeted them all warmly and looked as pretty as a picture to their fading eyes in her pink silk dress and white lace hat with a thread of pink ribbon for a band.

Among the guests was Charles Z. Gleason, ninety-two, a former White House policeman who had joined the staff during the McKinley administration and retired during the Hoover years. He wore his gold policeman's badge to identify his past association with the Executive Mansion.

Mrs. Eisenhower delighted in decorating the White House for the holidays—even Halloween—and it was not wholly because her four grandchildren (her only son's children) were frequent guests at the mansion. She loved to decorate, and thought their invited guests, as well as the thousands who toured the state rooms from ten to twelve o'clock five mornings a week, would enjoy them also.

She gave a luncheon for wives of White House staff members on October 30, 1958, and the guests arrived to find the usually dignified state rooms replete with goblins, witches, black cats, owls and ghost-white skeletons.

Eight spooky skeletons hung from wall lights in the State Dining Room; miniature witches on broomsticks perched on white tablecloths, and yellow jack-o-lanterns and shocks of corn stalks stood on the floor in the corners.

Swinging from the chandelier in the foyer were black cats (cut from black construction paper), disembodied witch heads, black owls and goblins, mixed with fall foliage. The white columns wore a facade of brown corn stalks, transforming the red-carpeted corridor into a harvest scene with huge real and fake pumpkins and rosy, red giant-size apples clustered at the base.

Place cards for the forty-nine guests were on the end of strings that trailed from yellow jack-o-lanterns spaced at intervals on the E-shaped table, alternating with huge silver bowls of bronze and yellow pompom chrysanthemums. At the other end of the strings were miniature goblins, ghosts, owls and cats with a fortune-telling note for each guest. Scattered over the white tablecloth were dried leaves, Indian corn, autumn nuts, squash and dried gourds.

Orange bulbs burned in the chandeliers, and Mrs. Eisenhower wore a gold printed silk dress and gold shoes to match the autumn decor.

It was a simple luncheon menu, that included

creamed chicken over asparagus, green salad, and frozen dessert in a yellow pumpkin mold.

Three weeks later, when Mrs. Eisenhower gave a Thanksgiving luncheon, early, for the forty-one women delegates and alternates to the United Nations, the E-shaped table was again decorated with the yellow pompoms in large silver urns, and autumn leaves, gourds, squash and Indian corn were scattered on the table. However, the menu was more elaborate and was typical of an American thanksgiving meal. It included:

Consommé Madrilène
Roast Turkey with Herb Dressing
Cranberry Sauce
Hot Rolls
Spinach Soufflé   Candied Sweet Potatoes
Shredded Lettuce with White Grapes and Blue Cheese
Pumpkin Chiffon Pie and Slices of Cheese

Mrs. Eisenhower sat next to the Russian delegate and explained to her the American customs of Thanksgiving. After luncheon, she gave the guests a tour of the White House.

This is Mrs. Eisenhower's variation of the traditional pumpkin pie:

PUMPKIN CHIFFON PIE

3 beaten egg yolks
3/4 cup brown sugar
1 1/2 cups cooked pumpkin
1/2 cup milk
1/2 teaspoon salt
1 teaspoon cinnamon
1/2 teaspoon nutmeg
1 envelope plain gelatin
1/4 cup cold water
3 stiffly-beaten egg whites
1/4 cup granulated sugar

Combine egg yolks, brown sugar, pumpkin, milk, salt and spice. Cook in double boiler until thick, stirring constantly. Soak gelatin in cold water, stir into hot mixture. Chill until partly set. Beat egg whites, add granulated sugar and beat stiff. Fold into gelatin mixture. Pour into baked pie shell and chill until set. Garnish with whipped cream. Makes one big pie.

Christmas was always a time for putting the mansion in holiday dress with Mrs. Eisenhower and the White House had never looked gayer at Christmas than it did during the years the Eisenhowers lived there.

There was always a huge tree in the East Room, brilliantly lighted, and dozens more trees were scattered throughout the mansion, upstairs, downstairs and even on the North Portico—and in every window hung a fat green wreath topped by a red bow, and over the front door, another very large wreath.

One year Mrs. Eisenhower placed a nativity scene, with figures nearly half life-size beneath the brilliant tree in the East Room, added a speaker that poured forth Christmas carols, and for her last Christmas in the mansion, she added a three-foot high Santa Claus in a red velvet costume to the collection beneath the tree.

The most beautiful tree that last Christmas (1960) however, was the tall spruce in the Blue Room with blue lights and blue Christmas balls. It served as a background for the President and Mrs. Eisenhower to receive White House employees at the annual Christmas party, at which they presented each employee a gift. Usually, it was a reproduction of a painting by President Eisenhower, and that Christmas it was a scene of a church in Bavaria silhouetted against fog-shrouded mountains.

The year before, Mrs. Eisenhower, wearing a red silk dress and diamond earrings, substituted for the President who was on a good-will mission abroad, at the traditional Christmas party for White House staff members. She stood in the East Room beside the ceiling-tall, glittering tree and shook hands with each, giving them a painting by the President. It was a picture of Mount Eisenhower, a peak in the Canadian Rockies near Alberta named in honor of the President.

In the foyer where the Marine Band played Christmas music, the white columns were entwined with green roping, and in the State Dining Room, the long table had holly leaves with red bows sprinkled over the white cloth. In the center of the table was a miniature train with its open cars filled with tiny Christmas-wrapped packages. Holly leaves circled the coffee urn and decorated the trays of doughnuts and coffeecake. On the mantel were miniature Santa Claus figures, and red carnations and greens tied with red ribbons decorated the wall lights.

Here are some of the holiday recipes Mrs. Eisenhower served:

SWEDISH SPRITZBAAKEN COOKIES

1 pound butter	2 whole eggs
1 cup sugar	2 extra yolks
4½ cups sifted flour	vanilla

Cream sugar and butter. Add beaten eggs, then flour gradually, then vanilla to taste. Use cookie press and form your own designs. Bake in moderate oven twenty minutes. Makes about one hundred cookies.

MILLION DOLLAR FUDGE

4½ cups of sugar
pinch of salt
2 tablespoons butter
1 tall can evaporated milk

Boil six minutes to form syrup.

Put into a large bowl:

12 ounces semi-sweet chocolate (chocolate bits)
12 ounces German sweet chocolate
1 pint marshmallow cream (2 jars)
2 cups nutmeats

Pour boiling syrup over ingredients in bowl; beat until chocolate is all melted and pour in pan. Let stand a few hours before cutting. Store in tin box.

Mrs. Eisenhower chose a St. Patrick's Day theme for the tea she gave for wives of members of the House of Representatives on March 12, 1959. This time the white columns in the foyer were twined with green ribbons and smilax festooned with miniature green top hats and green and white striped candy canes. Shamrocks hung by green ribbons from the chandelier. There were more shamrocks—green and silver foil ones—in the State Dining Room where flowers were green-tinted carnations and bells of Ireland. Two gold trays on the mantel were filled with green carnations and shamrocks hung from the chandelier and wall lights.

The food included caramel frosted cakes (a favorite of Mrs. Eisenhower's), open faced and cartwheel sandwiches, cream puffs filled with chicken salad, cookies, coffee, tea and grape punch. A harpist from the Marine Band played Irish airs as Mrs. Eisenhower, wearing an emerald green chiffon brocade dress, received the guests in the Green Room. The Marine Band played later, while the guests were in the dining room.

Many of the same decorations were used the following day when Mrs. Eisenhower gave an Irish luncheon for wives of members of the Cabinet and Senate. Added to the decor were eye-catching green-coated leprechauns popping out of flower pots that centered the nineteen tables. The tables were covered with white Irish linen cloths, and at each place was a corsage of green carnations with a tiny shamrock in the center.

The luncheon menu was:

Tomato Soup with Sour Cream
Sliced Turkey
Asparagus with Cream Sauce
Green Salad
Ice Cream
Coffee

Springtime was the theme of the luncheon she gave for Senate ladies on April 20, 1959. The State Dining Room was decorated with pink azaleas, pink snapdragons, pink gladioli and pink tulips. Pink tablecloths covered the round tables, and a real fountain was in the center of the room just opposite the First Lady's table. It was eight feet tall and surmounted by a gleaming white porcelain figure. Tulips bobbed and bubbled in the basin of the fountain.

Snapdragon petals floated in the fingerbowls, scented with lemon verbena. Out in the foyer, smilax and redbud twined around the marble columns and a pink flower cart stood just outside the door of the Blue Room.

Each guest received a pink engraving of the White House mounted on gold with her name underneath in matching pink Spencerian script.

Mrs. Eisenhower, in a black silk dress splashed with pink and red roses, greeted the guests in the Red Room. Standing with her was a former First Lady, Mrs. Woodrow Wilson.

On the menu were:

Cold Consomme
Breaded Veal
Lemon Mold Topped with Cherries
Heart-shaped Pink Cookies

The next year when Mrs. Eisenhower entertained the Senate wives, springtime was again the theme. This time, however, daisies—pink, white and yellow—decorated the State Dining Room. Guests dined on:

Chilled Melon with Prosciutto
Filet of Sole Amandine
Tiny Green Peas Beets
Hot Rolls
Lettuce Salad with Orange Sections
Ladyfingers with Whipped Cream and Candied Cherries

Her most elaborate party for Senate wives, however, was the luncheon on April 9, 1957, when the theme was birds, butterflies and bunnies. Masses of spring flowers—azaleas, hydrageas, tulips, iris and Easter lilies—banked the mantel in the dining room. On the mantel also were Easter bunnies hatching out of pale blue shells, and hanging from the chandelier and ceiling were giant gaily-colored net butterflies.

Yellow tulips, blue iris, yellow and white daisies in a vermeil bowl.

In the foyer, the white columns were entwined with real cherry blossoms and artificial roses while spring flowers filled the two niches on either side of the entrance to the Red Room.

In this festive atmosphere, birds sang liltingly and many guests thought they were real, although only two birds were visible swinging in a cage in the center of the ring at the bottom of the chandelier. Mrs. Eisenhower gave away the secret when she told an attendant to "turn down the birds." Then the guests knew the songs were recorded.

Each of the twenty tables was centered with a white vase of jonquils, iris and carnations and a tiny orchid-colored hatbox at each place contained a tiny Easter bonnet corsage for each guest.

On the menu were:

Salmon Mousse in Fish Mold
Lettuce Salad with Cheese Sticks
Cake with Maple Frosting
Fruit with Whipped Cream
Coffee

Mrs. Eisenhower gave a "farewell" luncheon on April 10, 1960, for women reporters who had covered her activities as First Lady. She gave them an insight into her view of life in the Executive Mansion. Even after nearly eight years in the White House, she said she still got a thrill out of giving a state dinner or a reception, and that her favorite room was the Red

Red and white carnations in an artistic arrangement center the table for a state luncheon at the White House in the 1960s.

Room. Not because it was more beautiful than the others, but because it was there she held her first reception, "and there is something special about a first," she said.

Mrs. Eisenhower received the women reporters in the Green Room, and then they went to the State Dining Room where masses of tangerine carnations, a new shade just being introduced, decorated the tables. A large silver "ship of state" was on the table directly in front of Mrs. Eisenhower. Guests drew numbers for places at the E-shaped table and dined off the Truman green-rimmed plates.

On the menu were:

Fresh Pineapple
Filet of Sole Amandine
Creole Rice — Asparagus
Endive Salad
Raspberry Mold with Currant Sauce

After luncheon the guests went to the East Room where they were joined by President Eisenhower to hear a short program by the Defiance College Choir, Defiance, Ohio. Under the direction of George Gerhardt

Apple and wild cherry blossoms with an assortment of Spring flowers in a pair of Sevres cachepots with gold bands that match the gold draperies at the windows and golden eagles which support the grand piano in the East Room.

Arkebauer, the choir sang "Battle Hymn of the Republic" and then "The Yellow Rose of Texas," one of the Texas-born President's favorite songs.

Mrs. Eisenhower's luncheons for wives of chiefs of state were more elaborate. At the luncheon on July 16, 1958, for Madame Chiang Kai Shek, the menu included:

Jellied Consommé Royal with Red Caviar and Sour Cream
Broiled Chicken with Rice
String Beans with Almonds
Green Salad with Roquefort Cheese
Raspberry Sherbet with Fresh Raspberries
Lady Fingers

At a luncheon in October, 1959, for Mrs. Lopez Mateos, wife of the President of Mexico, to which thirteen guests were invited, the menu included:

Cream of Almond Soup
Melba Toast
Hearts of Celery Assorted Olives
Broiled French Lamb Chops
(wearing pink paper frills)
Minted Fresh Pears Parsleyed New Potatoes
Thyme Peas
Orange and Roquefort Salad with French Dressing
Cheese Straws
Brandied Fruits
Nuts Coffee

Miniature white carnations, baby's breath, red sweet William, maroon bachelor's buttons, marguerite daisies and feverfew in an antique tole monteith beside a bust of Henry Clay.

Following are recipes for some of the luncheon favorites Mrs. Eisenhower served in the White House:

CHICKEN JEWEL SALAD RING

Cranberry layer:

1 envelope unflavored gelatin
1 cup cranberry juice cocktail
1 can (1 lb.) whole cranberry sauce
2 tablespoons lemon juice

Sprinkle gelatin on cranberry juice cocktail in saucepan to soften. Place over low heat, stirring constantly until gelatin is dissolved. Break up whole cranberry sauce; stir in gelatin mixture and lemon juice; turn into six-cup ring mold; chill until almost firm.

Chicken layer:

1 envelope unflavored gelatin
3/4 cup cold water
1 tablespoon soy sauce
1 cup mayonnaise
1 1/2 cups cooked diced chicken
1/2 cup diced celery
1/4 cup coarsely chopped almonds, toasted

Sprinkle gelatin on cold water in saucepan to soften. Place over low heat, stirring constantly until gelatin is dissolved. Remove from heat, stir in soy sauce, cool, then gradually stir in mayonnaise until blended. Mix in remaining ingredients. Spoon on top of the almost firm cranberry layer. Chill until firm. Unmold on salad greens.

DANISH TOMATOES

Slice tomatoes. Add dark brown sugar, a layer of tomatoes and a layer of sugar. Add vinegar, salt and pepper to taste. Put in refrigerator for a couple of hours before serving.

During his first four months in the White House President Eisenhower gave a series of what were termed "knife and fork" breakfasts, luncheons and dinners to which every member of Congress was invited, a few at a time. All except four attended, and they were intended as get-acquainted sessions. Throughout his two terms, he gave small stag dinners at intervals for business leaders, labor leaders, administration officials, Republican leaders, scientists, educators, editors, religious leaders and his old friends in the military. His purpose was to keep in touch with developing opinions in the many segments that make up American life.

President and Mrs. Eisenhower gave the traditional number of state dinners and receptions each year (except the years he was recuperating from his heart attack and other illnesses) and added others. Many of the extra dinners were for heads of state who formed an almost unending parade to the White House during the Eisenhower years.

First to receive the red-carpet welcome at a state dinner was President Jose Antonio Romon of Panama, and his wife, who came in September, 1953. Then followed King Paul and Queen Frederika of Greece, President Celal Bayar of Turkey, Emperor Haile Selassie of Ethiopia, President Giovanni Gronchi of Italy; President Sukarno of Indonesia, Queen Elizabeth II of England and her husband Prince Philip, who came in October, 1957, King Mohamed V of Morocco, Sir Winston Churchill, King Baudouin of Belgium, Princess Beatrix of the Netherlands, and Nikita S. Khruschev, chairman of the Council of Ministers of the U.S.S.R., and Mrs. Khruschev.

All were entertained at either a state dinner or state luncheon at the White House. If the events could be compared, the most brilliant probably would be the dinner for Queen Elizabeth and Prince Philip, and the most talked-about, the dinner for Soviet Premier Khrushchev and Mrs. Khruschev. Theirs was a formal dinner for some one hundred guests seated at the E-shaped table in the State Dining Room. It was a "white tie" event, as were all the Eisenhower state dinners, but the Premier showed up in a dark business suit with his decorations pinned to the jacket. Mrs. Khrushchev wore a short teal-blue gown with a diamond and emerald brooch at the neck. President Eisenhower wore white tie and tails, and Mrs. Eisenhower a long gown of gold brocade, diamond earrings and a necklace of diamonds and pearls.

The menu featured typical American fare: curry soup, roast turkey with cranberry sauce, sweet potatoes and tossed green salad.

The unusual and delicious soup was made as follows:

COLD CURRY SOUP

1/3 cup butter	1/8 teaspoon pepper
1/4 cup minced onion	1/4 cup flour
1/4 cup diced celery	1 quart milk
1 1/2 teaspoons curry powder	2 chicken bouillon cubes
1 teaspoon salt	flaked coconut

Melt butter in saucepan over low heat. Saute onion and celery in butter until transparent. Blend in seasonings and flour. Add milk, stirring constantly. Cook until smooth and thickened. Add bouillon cubes; stir until blended. Chill thoroughly. Pour into chilled bowls and sprinkle with flaked coconut. Makes six servings.

A few evenings later Premier and Mrs. Khrushchev gave a formal dinner in honor of President and Mrs. Eisenhower at the Soviet Embassy.

During their first social season, the Eisenhowers began the practice of inviting to each dinner two reporters who covered the White House to serve as "pool reporters" and give an account of the event to their colleagues. At the first dinner, wines and liquors were served, which turned out to be the big news of the evening, because the White House had announced previously that alcoholic beverages would be served only when required by protocol for the exchange of toasts.

That ended "pool reporters" coming in a working capacity. The two who had already been invited to the next dinner were notified that they would be welcome as dinner guests but not as reporters, which meant they were not to relate what they heard and saw at the dinner.

Thereafter, Mary Jane McCaffree, Mrs. Eisenhower's efficient social secretary, let society reporters come to the White House at seven-thirty before an eight o'clock dinner and view the table and decorations. Then they had to leave. They were allowed to return after dinner, however, and watch the arrival of additional guests invited for the after-dinner musical program, and could then stay for the program and the refreshments served later in the State Dining Room.

State dinners given by the Eisenhowers were as brilliant and elegant as any ever given in the White House but theirs also had a touch of sentiment.

At the dinner given for West German President Theodor Heuss on June 5, 1958, the Eisenhowers served a large portion of Americana. The menu included corn pudding, the after-dinner entertainment included a barber-shop quartet, and informality was represented by white dinner jackets instead of white tie.

Guests were seated at an E-shaped table decorated with high and low bouquets of purple iris, pink roses, pink and white snapdragons and the palest of yellow carnations. Green smilax feathered the white damask cloth. In the Red Room were great clusters of pink-fringed white peonies; in the Blue Room, yellow gladioli were banked against the windows; in the Green Room a profusion of pink gladioli; and in the East Room, masses of yellow carnations.

On the menu were:

Borscht
Small Pancakes filled with Creamed Crabmeat
Filet Mignon with Mushrooms
Corn Pudding Spinach
Fruit and Cheese Salad
(Jellied salad filled with black and red raspberries and creamed cheese balls)
Raspberry Sherbet Ring

Coffee and liqueurs were served in the Green Room to the men, and coffee for the ladies in the Red Room. The dinner guests were joined in the East Room by one hundred and fifty additional guests for a musical program by Meyer Davis and his orchestra and a barber-shop quartet called "The Ramblers."

A surprising addition to the program was a song titled "First Lady," written by Mrs. Meyer Davis and dedicated to Mrs. Eisenhower. The orchestra played favorite songs of the Eisenhowers' including "Tales from the Vienna Woods," "The Student Prince," "Serenade," "Lover Come Back to Me," and "Bless This House."

Two American folk numbers, "Aura Lee" and "Wish I Wuz," were sung by "The Ramblers" plus "Drink to Me Only With Thine Eyes," "The Old Songs," "Heart of My Heart," and "Coney Island Baby."

The chancel choir of the National Presbyterian Church, which the Eisenhowers heard nearly every Sunday when they went to church, provided the musical program following the dinner in June, 1960, in honor of Canadian Prime Minister and Mrs. John G. Diefenbaker. Mrs. Eisenhower missed the event because she was in the hospital. But she had a hand in planning it, including the menu that featured:

Chilled Melon with Prosciutto
Horseradish Soup
Crabmeat Creole Casserole
Brown Bread Sandwiches
Rock Cornish Game Hen with Wild Rice Dressing
Tomato Pudding String Beans
Hearts of Palm with Green Goddess Dressing
Vanilla Ice Cream Ring with Fresh Strawberries
Macaroons

Perhaps it was their western backgrounds (President Eisenhower was born in Texas and reared in Kansas, and Mrs. Eisenhower was born in Iowa and reared in Denver), but both had a warm friendliness that was quickly sensed by their guests. The President, who always looked out for Mrs. Eisenhower's health and well-being sometimes complained that she wanted to talk to everyone who came through the receiving line, which could be tiring and time-consuming, especially when there were several hundred guests to be received.

She devised a substitute for handshaking, however, when she welcomed 2,375 ladies of the Oriental Shrine to the White House on May 19, 1953. She stood on the new stairway in the foyer and said a few words to the ladies as they passed.

She liked to receive in the Red Room, and often had a group of ten or twelve women admitted at a time, and shook hands and greeted each. Then they passed through and another group would be ushered in and introduced to her. The procedure would be repeated until two or three hundred had been received in an afternoon.

Mrs. Eisenhower preferred to shake hands with her right glove off, and this caused a bit of social controversy in Washington, over whether a guest should wear her gloves or remove them to be received by the First Lady.

One woman, accustomed to wearing gloves when received by the two previous First Ladies, approached Mrs. Eisenhower at a reception and noticed her hand was ungloved. The guest quickly pulled off her glove and in so doing, off popped her bracelet which landed across the room. A White House aide went down on hands and knees and retrieved it, but it took much longer for the woman to regain her composure.

President and Mrs. Eisenhower celebrated their forty-third wedding anniversary in the White House on July 1, 1959. They invited two members of his official family who were celebrating wedding anniversaries at the same time, to join them in the party that featured a mock wedding. As the Marine Band played "Here Comes the Bride," Mrs. Eisenhower, wearing a bridal veil and carrying a bouquet, marched into the East Room, followed by Mrs. Neil McElroy, wife of the Secretary of Defense, who had been married thirty years, and Mrs. Leonard Heaton, wife of the Commanding Officer at Walter Reed Army Hospital, whose anniversary was June thirtieth. Both also wore veils and carried flowers.

After they got into the room where the three celebrating "bridegrooms" waited, the wedding music turned into a waltz and the couples took a turn around the floor. The President didn't dance long because of "a bum knee" from an old football injury sustained in his youth.

Just before leaving the White House the Eisenhowers gave a farewell dinner there for all the past and present members of his official family, plus a few close personal friends.

The off-white and gold state dining room with tables for ten set with the green-bordered Truman china. The green cushioned gilt chairs were purchased during the Kennedy Administration.

Determined to make it a nostalgic event, they invited the guests to their private quarters, the first time many of them had seen the family rooms, before going to the State Dining Room for dinner.

The carefully chosen menu included:

Minted Pineapple
Watercress Soup
Filet of Sole
Roast Stuffed Duckling served with Apple Sauce
Casserole of Wild Rice with Almonds
Braised Endive and Breadsticks
Bibb Lettuce Salad with Parmesan Cheese
and Green Goddess Dressing
Macaroon Charlotte

The U.S. Army Chorus sang "Auld Lang Syne" and "Bless This House." The U.S. Army Singing Strings played throughout the demitasse hour, and guests danced in the East Room to music by the Marine Orchestra.

Earlier that year, another dance was held in the East Room. This one was the "farewell" party John and Barbara Eisenhower, the President and First Lady's son and daughter-in-law, gave for their young friends. It was a black-tie dinner followed by dancing, the first time the Eisenhowers had turned the mansion over to the younger generation for entertaining.

Mrs. Eisenhower led a busy life in the White House in addition to entertaining. She directed the staff of seventy servants in keeping the household running smoothly and gave the furniture and woodwork the "white glove treatment" often to detect dust or dirt. She also completed the project begun by Mrs. Benjamin Harrison of collecting china representative of every presidential administration and had the collection displayed against a blue background in glass cabinets in the White House China Room.

The White House was the recipient of many historic acquisitions as a tribute to Mrs. Eisenhower. An anonymous donor gave a set of twenty-six gold-bordered plates with portraits of former First Ladies in the center painted by S. Wagner, a German artist. A large collection of vermeilware was acquired from the Margaret Biddle collection; and in June 1960, Mrs. Eisenhower accepted for the White House a roomful of antique American furnishings valued at $155,000 from the National Society of Interior Designers, Inc., of New York. There were fifteen pieces of American Hepplewhite and Sheraton, furnishing the small diplomatic reception room on the ground floor. It was the first time an entire room of furniture had been given to the Executive Mansion.

The Eisenhowers left the White House on June 20th, 1961, for their farmhouse near Gettysburg, Pennsylvania, which had been rebuilt and furnished during the years they were in the White House.

The Diplomatic Reception Room's Federal style furniture of New York and New England origin was presented to the White House during the Eisenhower Administration. The scenic wallpaper was added during the Kennedy Administration.

John F. Kennedy

chapter 24

A New Generation in the White House

Not since the days of Dolley Madison had the White House been the scene for such brilliant entertaining as was done by First Lady Jacqueline Kennedy and her history-conscious husband, John F. Kennedy—and not since the days of Thomas Jefferson, America's first gourmet of renown, had more serious thought been given to White House standards of food and drink.

A new generation had moved into the snow-covered executive mansion on that January 20, 1961, when John F. Kennedy, of Massachusetts, was sworn in as the 34th president of the United States. At forty-three he was the first President to be born in the twentieth century, and his beautiful, artistic wife, who was twelve years younger, was the nation's third youngest First Lady.

She had been living in the White House less than a week when she gave evidence she was not going to be an ordinary First Lady. In the months that followed she proved it.

There was hardly a facet of American life that did not feel the impact of her creative interests and imaginative originality. She brought new dimensions and stature to the position of First Lady, making it a job for accomplishments outside the realm of the President's duties.

One month and three days after moving into the White House, she announced plans to restore its state rooms with authentic furniture of the period when the house was built. She asked seven men and five women, all leaders in the field of arts, to serve on the Fine Arts Commission for the White House, which made a nationwide search for period furniture. During its first year, the Commission received hundreds of thousands of dollars worth of antique treasures for the mansion.

She focused on culture in all its aspects from the culinary arts to the performing arts. She gave encouragement to budding artists and focused attention on those who had already achieved.

She had a guide book to the White House published, emphasizing its history, and the number of visitors who entered its portals increased to more than a million a year.

She brought almost as many changes to the White House social functions as she did to the mansion's decor, and she set precedents with the food and entertainment served at official functions.

She began entertaining two days after moving into the White House and started by having her stepfather and mother, Mr. and Mrs. Hugh D. Auchincloss in for six o'clock tea. Two days later she had George Balanchine, the famous choreographer of the New York City Ballet, for tea, and they had a good time talking about ballet, an interest held by the First Lady since her childhood.

On a Sunday afternoon, nine days following the President's inauguration, the Kennedys gave their first official reception. It followed the swearing in of fifteen new presidential appointees and the guests who had been to White House parties in other administrations were instantly alert to the sweeping changes the Kennedys were making.

For the first time anyone could remember there was a bar in the State Dining Room, and butlers to stir up martinis, pour vodka, scotch, bourbon or champagne. There were ashtrays all around, indicating that guests could smoke if they wished. Previously, inexpensive glass ashtrays had been kept in table drawers and brought out for emergency use. About the only reminder of previous parties that afternoon was the Marine Band, playing from its traditional station in the foyer.

On the tea table in the State Dining Room were full-blown yellow tulips in silver bowls providing a forecast of spring to come.

President and Mrs. Kennedy came downstairs unannounced and through the Green Room to the doorway of the East Room where three hundred guests had assembled for the swearing-in ceremony. It was a friendly, informal atmosphere as they mingled with the guests, exchanging greetings.

Jacqueline Kennedy wore a black velvet sheath with a high boat neck in front cut to a shallow V in the back. Her pearl earrings barely showed under her swept-forward chestnut colored hair, and completing her jewelry were a diamond bracelet and diamond ring.

That afternoon she indicated that entertaining in the White House with her as hostess was going to be livelier and more exciting. This forecast was underscored eleven days later when she and the President gave the traditional Diplomatic Reception, to which the ninety-seven chiefs of missions and ten ambassadors from Latin American countries to the Organization of American States and their wives were invited.

The invitations read five to seven and there was a flurry of questions among diplomats as to what to wear. Traditionally the receptions had been held later in the evening and white tie was the proper dress. This time, most of them came in black tie.

They arrived to find fires burning brightly in every fireplace and the atmosphere that of a gracious but large private home. There was no receiving line. The President was in a jovial mood as he circulated through the crowd exchanging greetings, pausing here and there to have a quiet, serious conversation with some diplomat. He tried out his French on the French Ambassador and Madame Herve Alphand, asking only

"How are you?" before he lasped back into English to say proudly "My wife speaks very good French." Mrs. Kennedy was using not only French but her command of Italian and Spanish as she moved among the guests.

"You can tell a different person lives here," commented Mme. Alphand as she pointed out, first, the fires in the formerly empty fireplaces, then the lighted candles everywhere, and the flowers that were arranged the way the French women do them, natural and even a little wild, looking more as if they had come from the field than the florist.

The Kennedys gave their first state dinner on May 3, 1961, in honor of Tunisian President and Mrs. Bourguiba. Mrs. Kennedy had never before attended a state dinner at the White House but she had grown up in surroundings of formal entertaining and was not unfamiliar with the procedure. She made elaborate preparations for this one and went back into the records and read reports on state dinners that had been given by preceding Presidents and First Ladies. These reports provided a pattern and framework around which she applied her own originality. She spent hours in consultation with her capable social secretary, Letitia Baldrige, who had been a classmate at Miss Porter's School in Farmington, Connecticut. She received help and advice from the State Department Protocol Office. She conferred with the President, who took an active interest in the state dinners, and sometimes stopped by on the way from his office to the family quarters to sample the wines that were to be served at the dinner.

For the Tunisian visitors, Mrs. Kennedy selected a menu that would reflect the spring season and also be in keeping with the dietary customs of the honor guests. It consisted of:

Medallions of Cold Salmon
Roast Lamb with Vegetables
Salad and Brie Cheese
Molded Strawberry and Vanilla Ice Cream
Petits Fours

The guest list had been whittled down to ninety persons after much checking and rechecking of a master list originally compiled. Nineteen of the guests were members of the Tunisian President's cabinet or entourage and there was a similar representation from President Kennedy's cabinet and administration, plus a select group of outstanding citizens from various walks of life.

In making plans for the entertainment, Mrs. Kennedy had checked the weather forecast and was promised a clear night, but a little chilly. President and Mrs. Bourguiba arrived at the White House a few minutes before eight o'clock and had a brief visit with the Kennedys in the First Family's private quarters on the second floor of the mansion before the foursome descended the steps to the Marine Band's "Hail to the Chief."

Mrs. Kennedy wore a Grecian-style gown of pale yellow silk organza touched with brilliants. Mrs. Bourguiba, a handsome, older woman, wore a gown of blue-gray satin.

Following dinner, guests were directed outside to the South Lawn for the big surprise of the evening. It was a patriotic American military panorama staged on the South Lawn. Kleig-lights formed a stage for the parade of units from the Army, Navy, Air Force and Marine Corps in a setting that was dazzling.

The setting was a natural one, with the illuminated Washington Monument rising above the trees like a bright white sentinel to the left, and the South Lawn fountain playing in the center background. The shrubbery on the grounds served as curtains to shield each unit until its turn "on stage," so that the 480 performers seemed to appear from nowhere as they marched into the light. Overhead and in the distance could be seen the twinkling lights of planes coming in for a landing at National Airport, and visible on the lawn were the multicolored tulips blooming around the fountain.

President and Mrs. Kennedy and a dozen top-ranking guests watched from the second floor balcony outside the Blue Room. Mrs. Kennedy wore a mink coat over her sheer dress to ward off the 52 degree chill, and Mrs. Bourguiba wore a brocade coat collared in fur. On the terrace below sat the remaining eighty guests in chairs arranged theater-style.

Mrs. William Fulbright tried to share her mink stole with her husband, the Arkansas Senator, until military aides appeared with white blankets borrowed from the White House closets to wrap the shivering guests as they watched the brilliant performance.

The program which began just after ten o'clock ended on the dot of ten-thirty when both Presidents rose to take the final salute. President and Mrs. Kennedy, looking pleased at the apparent success of their first state dinner, escorted the state visitors back through the White House and bade them good-night at the door. The Kennedys remained at the door and shook hands with the other departing guests.

Simplicity, comfort and convenience were the guidelines Mrs. Kennedy followed in planning a state dinner, with a generous mixture of unpretentious glamor and excitement. She was always on the lookout for exciting new dishes to serve, and was an avid recipe collector. If she went to an embassy or restaurant and ate something she liked she would ask for the recipe.

President and Mrs. John F. Kennedy, left, and Vice President (later President) and Mrs. Lyndon B. Johnson make their entry at a White House party.

Two of her favorite dessert recipes, both elegantly beautiful and delicious, are shared here.

BOMBÉ GLACE MANDARIN

Enough ice cream or sherbet (orange or tangerine) to fill a mold
Sponge cake or lady fingers
Grand Marnier liqueur
Fresh oranges, tangerines, or banana slices

Spread ice cream inside mold to a thickness of about three-fourths inch. Spread one layer of diced sponge cake or lady fingers over ice cream. Moisten cake lightly with Grand Marnier. Fill rest of mold with ice cream. Put in freezer. When ice cream is firm, remove from mold and decorate with slices of fresh fruit.

STRAWBERRY MOUSSE

1 quart fresh or frozen strawberries
½ cup sugar
½ cup white wine
2 envelopes plain gelatin
½ cup cold water
½ cup boiling water
2 cups whipped cream

Select several choice strawberries for garnish and set aside. Press remaining berries through a fine sieve. Add sugar and white wine, stir well. Chill. Soften gelatin in cold water. Add boiling water and stir until dissolved. Cool. Combine gelatin and strawberry mixture, and beat until fluffy. Fold in whipped cream. Pour into a two quart oiled mold. Chill for three hours or longer. Unmold. Garnish with reserved strawberries. Serves eight.

While visiting in Greece one summer as First Lady, she enjoyed greatly the fish dishes there. When she returned to Washingron she asked the wife of the Greek Ambassador for recipes from the Embassy kitchen. One of these became a regular on the Kennedy's dinner menu. It was called Broiled Fish a la Grecque.

When she went with the President to a $1000-a-plate Democratic fund-raising dinner at Washington's new International Inn, she liked the Filet of Sole

Veronique served at the dinner and asked the hotel manager to have his chef send her the recipe. It is a fish served with white grape sauce.

One of her husband's favorite American recipes for fish, which he frequently enjoyed, was:

PRESIDENT KENNEDY'S FAVORITE FISH CHOWDER

2 pounds haddock
2 ounces salt pork, diced
2 onions, sliced
4 large potatoes, diced
1 cup chopped celery
1 bay leaf, crumbled
1 quart milk
2 tablespoons butter
1 teaspoon salt
pepper, freshly ground

Simmer haddock in two cups water for fifteen minutes. Drain. Reserve broth. Remove bones from fish. Sauté diced pork until crisp, remove and set aside. Sauté onions in pork fat until golden brown. Add fish, potatoes, celery, bay leaf, salt and pepper. Pour in fish broth plus enough boiling water to make three cups liquid. Simmer for thirty minutes. Add milk and butter and simmer for five minutes. Serve chowder sprinkled with diced pork. Serves six.

White House Chef Henri Haller and Kitchen Aide Julia A. Farrow prepare whole roast tenderloin of beef for a state dinner—a continuing White House scene.

Still another Kennedy favorite was:

NEW ENGLAND SEAFOOD CASSEROLE

1 pound canned or frozen crabmeat
1 pound shrimp, cooked, shelled and deveined
½ cup chopped green pepper
¼ cup finely chopped onion
1½ cups finely chopped celery
½ teaspoon salt
1 cup mayonnaise
1 tablespoon Worcestershire sauce
2 cups crushed potato chips
Paprika

Combine crabmeat, shrimp, mayonnaise, green peppers, onion, celery, salt, mayonnaise and Worcestershire. Pour into a buttered casserole. Top with crushed potato chips and sprinkle with paprika. Bake twenty-five minutes in pre-heated 400 degree oven. Serves eight.

Mrs. Kennedy had a distinct preference for French cooking, and soon after moving into the White House, began looking for a chef to add the inimitable French touch.

She subsequently lured Rene Verdon, the French expert, away from the Carlyle Hotel in New York to preside over the White House kitchens. The French flavor he added to state dinners was an instant success. "People talk about what Mrs. Kennedy has done redecorating and restoring the White House, but they should talk more about how she has improved the food," Mrs. Alben W. Barkley, widow of the former Vice President, commented after attending one of the Kennedy's state dinners.

In planning a menu, Mrs. Kennedy kept an eye on cost and tried to select foods that were in plentiful supply and in season. She also kept the number of courses to four although the previous White House dinners ran to five and six, and during the Grant administration ran as high as twenty-nine.

Mrs. Kennedy's streamlined menus consisted of either a fish or soup course first, then the main dish, salad and dessert. She followed the same menu pattern for her family meals, except that then, she usually eliminated dessert and ended with salad and cheese.

Recipes for dishes served to both family and friends were:

CONSOMME JULIENNE

1 tablespoon butter (or substitute)
2 medium-size carrots
1 stalk celery
1 small turnip
1 leek
1 small white onion, thinly sliced
½ cup finely-shredded cabbage
½ teaspoon salt
dash of pepper
1 teaspoon sugar
4 chicken bouillon cubes
4 cups boiling water

Slice carrots, celery, turnip and leek into thin two-inch strips and add with cabbage, onions, salt, pepper and sugar to melted butter in a saucepan. Cover and cook over low heat for about five minutes. Dissolve bouillon cubes in boiling water and add vegetables. Simmer five minutes. Garnish with chopped parsley. Serve with cheese straws. Serves six.

Pink and blue flowers in a vermeil bowl.

PRESIDENT KENNEDY'S FAVORITE BEEF STROGANOFF

2 pounds beef
1 tablespoon flour
2 tablespoons butter
2 cups beef stock
2 tablespoons heavy sour cream
2 tablespoons tomato juice or paste
3 tablespoons grated onion
butter to sauté

Cut beef in thin strips, sprinkle freely with salt and pepper, and let stand for two hours in cool place.

Make a "roux" by blending flour with butter over gentle heat until mixture bubbles and is smooth. Gradually stir in beef stock and cook until mixture begins to thicken. Boil for two minutes, then strain into a saucepan. Add heavy sour cream alternately with two tablespoons tomato juice or paste, stirring constantly. Simmer very gently, without boiling.

Brown the pieces of beef in three tablespoons butter with grated onion. When the meat is brown, pour the meat, onion, and butter into the sauce, taste for seasoning, and simmer gently, or cook in a double boiler over hot water, for twenty minutes.

She turned her menus over to the executive housekeeper for the mansion, who did the buying for official functions as well as private meals. Ann Lincoln, who became the Kennedys' executive housekeeper, usually did the marketing once a week, telephoning the order to Henry Brothers, a small independent grocery store a few blocks from the White House.

Mrs. Kennedy personally selected the china used at state dinners, and her favorite was that purchased during the Benjamin Harrison Administration. Designed by Mrs. Harrison, it has a cornsheaf and goldenrod rim and in the center of the plates an eagle, wings outspread, perched on the shield of the United States.

Because there were not enough settings of the Harrison china for large dinners, however, Mrs. Kennedy used the Truman and Eisenhower china more often.

To keep the table as uncluttered and simple as possible, she did not use bread and butter plates—following instead the French formal dinner settings.

Folded white napkins were placed on the plates, and gilt-edged place cards laid on top of the napkins. At each place was an urn of cigarettes, a salt dish with a tiny spoon, an ashtray with White House matches, and an individual nut dish.

The table service, which she kept to only what was actually needed, was not silver but vermeil, and two of the knives had gleaming opalescent pearl handles.

Four long-stemmed tulip-shaped glasses were set beside each plate. One was for water, two for wine, and the fourth for the champagne for the toasts.

When Mrs. Kennedy and the President dined privately in their family quarters on the second floor, it was usually by candlelight, and they had no music, for it was then they kept each other informed on current plans and projects.

Mrs. Kennedy chose the entertainment for the White House as carefully as she did the food. She felt that entertainment presented at the White House should represent the best in American life and she was delighted if she could give recognition to some undiscovered or unrecognized talent. When she read in a national magazine about a young mezzo-soprano from St. Louis, Grace Brumby, who had won honors at the Bayreuth Festival while practically unknown in her own country, the First Lady asked her social secretary to check on the singer's talent and schedule, to see if she would sing at the White House.

The twenty-four-year-old Negro was receptive, and came to sing at a dinner jointly honoring the Vice President, the Speaker of the House and the Chief Justice of the Supreme Court. She sang for the White House audience "O Don Fatale" from Verdi's opera Don Carlos, the same number that won her acclaim at the age of seventeen when she sang it on Arthur Godfrey's radio-television Talent Scout program in 1954.

Her selections at the White House included also two eighteenth century Italian love songs, two numbers by Richard Strauss, two by Duparc, an American Folk Song, "Boatmen's Dance" by Aaron Copeland, and an American spiritual.

At a dinner on April 29, 1962, honoring Nobel Prize winners in the Western Hemisphere, the White House staged a "coup" for the blue-ribbon audience.

It was three readings by actor Frederic March. One was an excerpt from one of the late Ernest Hemingway's unpublished works dug out of the vaults by his widow; the others were the preface to "Main Street," the novel for which Sinclair Lewis won the Nobel Prize in 1930; and four paragraphs from the speech in which General George Catlett Marshall launched his plan for helping Western European nations recover from the devastation of World War II, and won his Nobel Prize in 1953.

There were forty-nine Nobel Prize winners among the 175 guests and it was the largest number at a White House dinner in recent years. They overflowed from the fourteen round tables in the State Dining Room to five tables in the Blue Room.

Another dinner that turned the White House into a brilliant international culture center was the one the Kennedys gave to honor one of France's most distinguished men of letters, Andre Malraux, the sixty-year-old Minister of State for Cultural Affairs for the Republic of France.

The guests, all distinguished in the arts, were served an evening of music, including Schubert's "Trio in B Flat Major, Opus 99" in four parts played by Eugene Istomin, pianist; Isaac Stern, violinist, and Leonard Rose, cellist.

The spring menu was impressive and international. It consisted of:

Consommé Madrilène Iranien
Lobster en Bellevue
Stuffed Bar Polignac
Potatoes Parisienne   Florida Asparagus
Pheasant Aspic Salad
Cream Puffs garnished with Nuts

The guests were seated at twelve tables in the State Dining Room and five tables in the Blue Room, and both rooms were an old-fashioned garden scene. Natural-looking arrangements of lilies of the valley, baby's breath, red and white tulips, blue iris, daisies, lavender, pink sweet peas, violets and sand lilies centered each table.

The all-American western ballet, "Billy, the Kid," was presented on a portable velvet-mounted stage in the

East Room by the American Ballet Theater of New York, as the entertaining climax to the dinner in honor of the President of the Republic of the Ivory Coast and Madame Felix Houphouet-Boigny. And when the Grand Duchess Charlotte of Luxembourg was entertained at the White House, Mrs. Kennedy added a special reading dedicated to President Kennedy to the program of Elizabethan poetry and music presented by actor Basil Rathbone and the Consort Players, a scholarly group of musicians who played sixteenth and seventeenth century instruments.

The special reading by Rathbone was "Henry V's" famed "St. Crispin's Day Speech" spoken on the eve of the Battle of Agincourt, and was added because President Kennedy was so fond of quoting those lines by Shakespeare himself.

The White House took on an opening-night-on-Broadway atmosphere at the dinner honoring the Shah and Empress of Iran. The entertainment was Jerome Robbins' "Ballets: U.S.A." which Mrs. Kennedy had seen both in Europe and New York and enjoyed so much. The company had been disbanded, and some of the fifteen original dancers were in other Broadway musicals, but were released from those shows for one night for the performance at the White House. Electricians, carpenters, stage directors and stagehands had worked around the clock to prepare for this special show.

After the dinner honoring the Sudan's English-speaking President Abboud, the players from the American Shakespeare Festival Theater of Stratford, Connecticut did the prologue from "Henry V," the Macbeth murder scene, a love scene from "Troilus and Cressida," "Seven Ages of Man" from "As You Like It," and Prospero's speech from "The Tempest." It was a full evening of Shakespearean drama, and among the guests, attending their first state dinner, were the two teen-aged daughters of Vice President and Mrs. Lyndon B. Johnson.

When Lynda Bird and Luci Johnson first received their invitations, they thought a mistake had been made, and telephoned Mrs. Kennedy to ask if she realized they were only seventeen and fourteen. She said she did, and wanted them to come and enjoy the Shakespeare players.

Mrs. Johnson's advice to her daughters before going was "Don't drink any of the wines at the White House."

She knew the finest wines were served and that they could be heady to teen-aged drinkers. One of President Kennedy's state visitors at another dinner found the wines heady also, especially on top of drinks he had imbibed before arriving at the White House.

Before the dinner ended, this guest of honor began to show signs of intoxication. President Kennedy recognized the situation, speeded the dinner to an end, and led his guest of honor to the library on the ground floor of the White House. President Kennedy was presenting his distinguished visitor a copy of his book, "Profiles in Courage" when the man suddenly dozed off in mid-sentence. He was helped into a waiting limousine that had been summoned and, after a brief ride in the chilling night air, he was returned to the privacy of Blair House (the President's guest house) where he was staying.

Other guests were not aware that the guest of honor was not in his front row seat during the after-dinner entertainment in the East Room that evening.

There were many other notable social events during the swiftly-passing three years the Kennedys occupied the White House. Among them was the concert given by cellist Pablo Casals following a dinner attended by the foremost American composers and conductors. Among the guests was Alice Roosevelt Longworth, daughter of former President Theodore Roosevelt, who had heard Casals play at the White House in 1904, when her father was President.

The Kennedys gave many private parties that were as brilliant as their state dinners. There was a special dinner for the great Russian-born composer and conductor, Igor Stravinsky, and guests came from London, Paris, New York and Chicago. The eighty-year-old composer did not play any of his compositions on the grand piano in the East Room, but a Kennedy staffer who had been a child prodigy at the piano gave a whimsical performance after Stravinsky left. It was Pierre Salinger, the President's press secretary, who played a composition he wrote when he was fifteen years old.

At another dinner in honor of former President Harry S. Truman, also a piano player, he turned star performer after dinner and played for the forty-three guests following a concert by Eugene List.

It was not unusual for guests invited to President and Mrs. Kennedy's state dinners and private parties to see a wide-eyed little girl with blond hair sitting in pink pajamas and robe on the top of the stairs, watching the parade of guests. Occasionally the President and First Lady would look up and smile, and she would wave to them. It was their four-year-old daughter, Caroline, who was not too young to enjoy from her vantage point on the steps the excitement that was going on in the house of history.

Of all the brilliant social events the Kennedys gave in the White House none outranked the state dinner they gave on the lawn at Mount Vernon, home of America's first president, in honor of the President of Pakistan, Ayub Khan. A huge tent was erected on the lawn high above the Potomac River, and sixteen round tables covered with yellow damask were set up underneath the colorful marquee to seat the 138 guests.

The neatly manicured grounds were sprayed three times that day with insecticides, the last time less than an hour before the time for guests to arrive, to free the area from gnats and mosquitos. As a further precaution against the insects, citronella candles burned on the tables.

Guests were transported down the Potomac River on the President's three yachts, and on a fourth one borrowed from the Secretary of Navy. Music for dancing was provided on the trip down the river and back.

On arrival, the guests were treated to a parade of the Fife and Drum Corps, whose members were dressed in Colonial uniforms patterned after those worn by the Continental Army at the time of the American Revolution.

Before dinner, the guests were given a tour of the home of George Washington, a national shrine owned and maintained by the Mount Vernon Ladies Association of the Union. The menu was one that could be prepared at the White House and transported in military vehicles that served as field kitchens, the twelve miles to Mount Vernon. It included:

Avocado and Crabmeat Mimosa
Poulet Chasseur
Couronne de Riz Clamart
Framboises à la Creme Chantilly
Petits Fours Secs
Demitasse

The wines were Haut-Brion Blanc 1958, Moet et Chandon Imperial Brut 1955, and liquers after-dinner.

After-dinner entertainment was a concert by the National Symphony Orchestra with Lloyd Geisler conducting. Selections played were Gould's "American Salute"; Mozart's "Allegro con spirito from Symphony No. 35 in D Major," "Haffner"; Debussy's "Dialog Between the Wind and the Sea" from "La Mer"; and Gershwin's "American in Paris."

Mrs. Kennedy introduced the practice of mixing men and women guests at coffee after state dinners. Formerly the men went with the President to the Green Room for coffee and cigars, and the ladies went with the First Lady to the Red Room for coffee. The Kennedys let the guests go to whichever room they chose.

Mrs. Kennedy was the first President's lady to travel abroad on her own while her husband occupied the presidency. As a goodwill ambassador abroad, she ranked at the top among all women in American history. She was such a hit in Paris and Vienna on her visit there with President Kennedy in the summer of 1961, that he told one audience "I do not feel that it is inappropriate for me to introduce myself. I am the man who accompanied Jacqueline Kennedy into Paris."

U.S. Ambassador to India Kenneth Galbraith requested that her influence for goodwill be extended to that country by a personal visit, and one was scheduled.

Mrs. Kennedy's one big problem in the White House was the lack of privacy for her children, Caroline, four, and John Jr., who was only two months old when he moved into the mansion. "How can I bring up normal children if they are to be treated that way?" she asked as photographers connived to get photographs of the children at play and on their daily outings.

Mrs. Kennedy started a nursery school in the White House for Caroline and her specially-chosen classmates. When John, Jr., got older, a class was added for him.

No challenge in the White House was too great for Jacqueline Kennedy, who had proven her resourcefulness as an inquiring photographer for a newspaper before her marriage to the handsome eligible Senator from Massachusetts who was destined to become President.

Both had been born to wealth, but Jacqueline had also been surrounded by culture and standards that were decidedly highbrow in comparison to those of her husband, who selected instead a career in politics and public service. However, he was guided by her to participation in the aesthetics and had the same high appreciation as she for talent and achievement in the world of culture.

During their three years in the White House, an era that ended when an assassin's bullet cut short the President's life on November 22, 1963, Mrs. Kennedy established a pattern for entertaining that enriched the social history of the mansion and provided a challenge for her successor to do as well.

The following are some of her menus and recipes that may be used for public and private entertaining elsewhere:

Luncheon in Honor of
the Prime Minister of Greece and
Mrs. Caramanlis
April 1, 1961

Consommé Ambassadeur
Chicken LaFayette
White House Terrine
Dame Blanche Pastry
Demitasse
Wines: Puligny-Montrachet Pucelles 1958
Corton Grancey 1955
Dom Perignon 1952

Luncheon for Senate Ladies
May 23, 1961

Demoiselles de l'Ocean
en Bellevue
Supreme of Capon Demidoff

Potatoes Parisienne
Green Beans with Almonds
Pineapple Sherbet
Petits Fours
Demitasse
Wines: Almaden Grenache Rosé

Luncheon in Honor of Prince Rainier
and Princess Grace of Monaco
May 24, 1961

Soft Shell Crab Amandine
Spring Lamb à la Broche aux Primeurs
Salade Mimosa
Strawberries Romanoff
Petits Fours Sec
Demitasse

Luncheon in Honor of the Vice President
of the Republic of China and Mrs. Chen
July 31, 1961

Melon and Ham
Chicken in Wine
Green Peas   Potatoes Rissole
Mixed Green Salad
Cheese and Crackers
Baked Alaska
Demitasse

Luncheon in honor of the President
of Finland and Mrs. Kekkonen
October 16, 1961

Truites sux Amandes
Poulet a l'Estragon
Epinards à la Crème
Tomatoes Grillé   Pommes Dauphine
Salade et Fromage
Tarte Maxime

Dinner for the Prime Minister of India
Jawaharlal Nehru
November 7, 1961

Vol-au-Vent Maryland
Gigot d'Agneau aux Flageolets
Tomatoes Grillé
Epinards à la Crème
Mousse aux Concombres

Bombé Glace
Caribienne
Wines: Almaden Pinot Blanc
Chateau Haut Brion 1953
Cuveè Dom Perignon Brut 1952

Dinner in Honor of the President of the Ivory Coast and Mrs. Houphouet Boigny
May 22, 1961

Quenelles de Sole au Gratin
Roast Sirloin
French Peas   Potatoes Gaufrettes
Galantine of Chicken
Green Salad
Bombé Glacés Pistachio
Petits Fours

Luncheon in Honor of the Prime Minister of Italy Amintore Fanfani
January 16, 1963

Braised Rockfish Daumont
Roast Lamb Dauphine
String Beans   Salade Mimosa
Savarin Almina
Demitasse

President Lyndon B. Johnson joins Santa Claus at a Christmas party in the White House for children from Washington area institutions.

Lyndon B. Johnson

chapter 25

THE LIVELY LBJs

White House entertaining continued with verve and imagination after Lyndon B. Johnson and his petite, charming wife, Lady Bird, moved into the mansion as President and First Lady in December, 1963.

Following so closely the sophisticated, artistic style of the Kennedys and the warm, lavish hospitality of the Eisenhowers, it might have seemed impossible for the Johnsons to set any new record in entertaining. But they did.

Naturally friendly and gregarious, the Texas-born President and First Lady brought a warm, informal atmosphere to the White House that made even formal functions easy and enjoyable. Under the creative coordination of Social Secretary Bess Clements Abell, daughter of a former Kentucky Governor, White House entertaining by the Johnsons hit new heights of elegance and hospitality.

It also established precedents in types, forms and reasons for presidential party-giving.

There were many innovations in the Johnsons' pattern of entertaining. They were the first to give a state dinner in the Rose Garden, a beautiful rectangular spot bordered by trees and flowers just outside the President's office.

Held on a perfect summer evening in June, 1964, it was in honor of their old friend, Dr. Ludwig Erhard, Chancellor of the Federal Republic of Germany, who had been entertained six months earlier at the LBJ Ranch in Texas.

One of the most original and brilliant of all the dinners given by the Johnsons, plans for it were made in duplicate. One plan called for a setting in the Rose Garden. The alternate plan involved holding the dinner inside the mansion in event of rain. Even in choosing the china for the tables set up in the Rose Garden, an equal number of servings were earmarked for use in the State Dining Room in the event a cloud appeared in the sky.

But it was a clear, sparkling evening and the weather was as if made to order. Japanese lanterns glowed over the flower-decked tables, stars twinkled in the sky and midway in the dinner, a full June moon rose over the trees in the garden casting moonglow on the brilliant setting.

The Marine Band was stationed on a flagstone terrace at one end of the garden, and played lilting tunes as guests found their places at the tables.

President and Mrs. Johnson received them in the East Room, from which they walked through the Green and Blue Rooms, out through the latter's French doors onto the South Balcony, down the steps into the garden. Thus, those attending their first state dinner had a chance to see the inside of the mansion also.

After dinner, guests returned to the ground floor corridor and rooms for coffee and liqueurs, and then wandered back onto the South Lawn, where chairs had been arranged before a half-shell bandstand on which the evening's program was presented. It featured music by the National Symphony Orchestra, and a ballet program by Maria Tallchief, prima ballerina, and Jacques D'Ambroise, principal dancer with the New York City Ballet.

Champagne and other refreshments were served afterwards from gaily-striped tents on the lawn. Guests stayed and strolled the grounds as long as they wished, enjoying the beauty of the event, so like a mid-summer night's dream.

A year and two days later, on June 14, 1965, President and Mrs. Johnson staged another innovation in the same setting. It was the first White House Festival of the Arts, which offered culture in some form for everyone. More than a year in the planning, the events began at ten o'clock in the morning and continued until midnight without a noticeable flaw.

A collection of contemporary American work in sculpture was exhibited in the East Garden (dedicated earlier by Mrs. Johnson as the Jacqueline Kennedy Garden in honor of the former First Lady). Exhibits of painting and photography were held inside the mansion in the East Wing corridors. Programs of poetry and prose, drama, ballet, symphony and jazz were presented at intervals throughout the day and evening. These were presented on a stage in the East Room during the day, and on the half-shell bandstand on the South Lawn in the evening.

At twilight the President and Mrs. Johnson gave an outdoor buffet supper on the South Lawn for four hundred guests, who dined at candlelit tables under the historic old trees. The guest list was a roster of "who's who" in the world of American arts and they came from all over the country. At the dinner, Mrs. Johnson sat at a table between actor Gene Kelly and photographer Edward Steichen.

At the prose and poetry presentation, Mark Van Doren served as narrator for readings by authors Saul Bellows, Catherine Drinker Bowen, John Hersey, and poetess Phyllis McGinley.

At the music presentation, Marian Anderson was narrator for selections by the Louisville Orchestra from Kentucky, and selections by Metropolitan Opera star Roberta Peters. For the drama segments, Helen Hayes narrated for selections from Tennessee William's "The Glass Menagerie," Frank Gilroy's "The Subject Was

Roses"; Arthur Miller's "Death of a Salesman"; and Millard Lampbell's "Hard Travelin'."

Actor Charles Heston was narrator for the afternoon screening in the State Dining Room of scenes from five motion pictures including "North by Northwest," "On the Waterfront," "Shane," "The Best Years of Our Lives," and "High Noon." Actor Gene Kelly introduced the dance program by the Robert Joffrey Ballet, and the Jazz program by Duke Ellington in the evening.

It was a national festival honoring contemporary American achievement in the arts, and nothing like it had ever been staged at the White House before. It set a pattern that could be followed at state, city and community levels giving recognition to local achievement in the arts.

This was only the beginning of the parade of culture the Johnsons brought to the White House. They staged more cultural and artistic events in more unusual settings and circumstances than had any of their predecessors in the White House.

Entertainment received equal billing with food at all the Johnsons' social functions. The former ran the gamut from hootenany singing to grand opera; from Indian dancers to national ballet; from University players to Broadway productions.

At their first state dinner, given a little more than a month after moving into the White House, the Johnsons presented Robert Merrill, baritone of the Metropolitan Opera, singing arias from "La Traviata" and "The Barber of Seville." On the same program appeared the New Christy Minstrels, nine young folksingers accompanied by guitars. They sang "Saints' Train" (a hootenany version of "When the Saints Go Marching In") and "Cotton Picker."

The guest of honor was Italy's seventy-two-year-old President Segni, who tapped his foot on the floor in rhythm to the music.

At this, as at subsequent state dinners given by the Johnsons, the guest list was an interesting collection of judges, labor leaders, musicians, actors, editors, scientists, lawyers, politicians, bankers, business executives, and personal friends.

 The Johnsons brought "Indians" to the East Room

Strolling violinist (right) provides music during a state dinner given by President and Mrs. Lyndon B. Johnson.

Secretary of Defense and Mrs. Robert S. McNamara arrive at the entrance of the Diplomatic Reception Room for a state dinner.

stage, at a dinner for President Maurice Yameogo of Upper Volta in March, 1965, when students from Santa Fe gave a performance of Gahan (mountain spirits revered by the Apache), wielding symbols of lightning.

In choosing entertainment for state visitors, Bess Abell tried to select something that would be representative of the best of American talent and at the same time something the guest of honor would enjoy. Knowing King Hassan of Morocco had a preference for ballet or Spanish flamenco, she invited Mexican-born dancer-choreographer Jose Limon to stage selections from "The Moor's Pavane" at the dinner.

Following the dinner for Great Britains' Prime Minister and Mrs. Harold Wilson, a forty-minute "Salute to American Musical Theater' was staged on the South Lawn. First presented a month earlier in New York, it was a revue of musical hits of the last one hundred years, and was staged by the Manhattan School of Music with David Wayne as narrator.

A whimsical musical, "The Voyage to the Moon," by Jules Verne, was presented by the Opera Company of Boston at the dinner honoring Turkey's President and Mrs. Cevdet Sunay, and as a special tribute to the visiting leader, part of the lyrics were translated and sung in Turkish.

A dinner honoring the Vice President, the Speaker of the House of Representatives and the Chief Justice, featured Carol Channing, and a cast from the Broadway hit produced by David Merrick, "Hello Dolly." Excerpts were presented on the East Room stage, after which President and Mrs. Johnson produced an unexpected surprise. It was a huge birthday cake on which flickered four candles, marking the fourth anniversary of the show's run on Broadway.

With teen-aged Luci Johnson, and her older sister, Lynda Bird, who had just turned twenty, in the White House there were many parties for young people and children. The Boston Symphony Orchestra under the direction of Eric Leinsdorf played a concert for teen-aged children from Embassy Row.

The National Players presented the "wooing" and "wedding" scenes from "The Taming of the Shrew" by William Shakespeare one afternoon, for one hundred children of members of Congress, and Luci's sophomore

The East Room transformed into an outdoor garden for a dinner dance given by Lynda Bird Johnson for Princess Irene of Greece.

Dancing at the White House in the 1960s.

Several of their holiday recipes are printed below.
class at National Cathedral School. That same spring, Luci gave a reception for the junior and freshman classes at the school on a Sunday afternoon. David Bean, a young local pianist, was invited to play.

When Luci and Lynda sent an invitation, it was on a card with an engraving of the White House in black and white at the top, in place of the gold seal that marked the President's invitations.

One of the prettiest parties ever given at the White House was the dinner-dance Lynda Bird gave in February, 1967, in honor of Princess Irene of Greece.

Although the weather was cold and icy outside, the East Room was transformed into a festive garden setting. Green topiary trees decorated with pink paper flowers and tiny bee lights formed a ring around the East Room. Between the tables was a framework of bare branches upon which white Japanese lanterns were hung. Tables, each seating ten, were covered with pink cloths and decorated with "bamboo" vermeil bowls of spring flowers. The green-bordered Truman china and vermeil flatware added color to the table. (Menu at the end of chapter.)

Music for dancing between dinner courses and after dinner was provided by the swinging Lee Evans Orchestra from New York City. Lynda and her guests danced the frug, swim, monkey, watusi and others popular with the young set.

Log fires burned brightly in the fireplace in the Green Room, where Lynda and the slender young Princess received nearly two hundred of the guests. When the girls sat down at their table, they found a single red rose at each place. The roses added a romantic note to the occasion, for they were part of the three hundred sixty-four red roses Lynda Bird had received the day after Valentine's Day from her favorite beau, Actor George Hamilton.

Hamilton, in North Africa at the time, sent a single long-stemmed rose to Lynda on Valentine's Day with a note: "You are my Valentine today and everyday of the year." The next day 364 red roses arrived from George.

At her first White House party two years earlier, Lynda had asked guests to come in masquerade as figures in American history. She herself came as a

Southern belle dressed in a flowing white hoop-skirted dress decorated with bright silk ribbons and a garland of yellow roses around the off-shoulder neckline. Some of the young men came in costumes of the American Revolution, some as Mississippi riverboat gamblers, some as gunmen of the old West, and one as a much-decorated pilot of World War I, in his flying boots and jacket.

At Christmas time each year the Johnsons gave a series of parties for children and adults. One party each Yule season was for children of White House staff members, and the entertainment was always bright and colorful. One year it was Bavarian folk dancers; another year it was dancing by costumed members of the Alliand College Kujawiak, a group organized to preserve Polish Folk Art. They presented a Polish wedding dance, and the dancers tossed paper flowers into the audience. One year the children of the diplomatic community were invited to a party featuring a performance by the Old Guard Fife and Drum Corps in their white wigs and red colonial costumes. The Army musicians led the boys and girls from many lands dressed in their native costumes, on a march around the Christmas tree and into the State Dining Room for pink punch and Christmas cookies.

President and Mrs. Lyndon B. Johnson and Vice President and Mrs. Hubert H. Humphrey, right, join Broadway Star Carol Channing in celebrating the fourth birthday of "Hello Dolly" after she presented excerpts from the musical at the White House following a state dinner.

TUILES WITH ALMONDS

½ cup sugar
2 large egg whites
½ cup sifted flour
½ teaspoon almond extract
½ teaspoon pure vanilla extract
¼ cup butter
½ cup blanched almonds finely shredded
Confectioners' sugar

Beat the sugar with the egg whites until the egg whites are frothy. Stir in the next five ingredients. Mix well. Drop the batter from a teaspoon in mounds the size of a quarter onto lightly buttered baking sheets, or pipe through a pastry bag onto lightly buttered baking sheets, leaving two inches between cookies to allow room for spreading. Sprinkle tops with confectioners' sugar. Bake the cookies in a preheated oven (350°) eight minutes, or until they have browned lightly around the edges. Cool one-half minute on the baking sheets, then remove, and while the cookies are still hot, bend them around a small rolling pin or small glass to give them the shape of a curved roofing tile. Store cookies in tightly closed container. Makes about two and one-half dozen cookies.

Japanese lanterns on bamboo poles provide shadowy lighting for a dance on the roof terrace.

ALMOND POTATO CROQUETTES

2 pounds potatoes
1 teaspoon salt
¼ teaspoon ground white pepper
3 whole eggs and 2 egg yolks
Flour
Blanched almond shavings

Peel and quarter potatoes. Cook until soft in a covered saucepan with one inch boiling water and the salt. Drain potatoes well and put them through a ricer or food mill. Add pepper. Beat two of the whole egg yolks together until light and foamy and add to potatoes. Whip the mixture until fluffy. Sprinkle a pastry board lightly with flour and shape the potatoes into croquettes or balls. Beat the remaining egg slightly. Dip croquettes in beaten egg and then roll them in thin blanched almond shavings. Fry in deep fat preheated to 375 degrees until browned. Drain on paper towels.

A treasured family recipe is for:

PRESIDENT JOHNSON'S FAVORITE DOUBLE DIVINITY FUDGE

2 cups sugar
2/3 cup water
1/2 cup light corn syrup
2 egg whites, slightly beaten
1 teaspoon vanilla
dash of salt (added to egg whites)

Combine 1/4 cup sugar and 1/3 cup water; cook until small amount of syrup forms a soft ball in cold water (240°). Separately, cook remaining 1 1/2 cups sugar, 1/3 cup water and corn syrup until it forms a hard ball in cold water (250°). Let first syrup cool slightly. Add slowly to egg whites, beating constantly about one to two minutes or until mixture losses its gloss; add second syrup in the same way. Add vanilla, and turn into greased pan. Cut in squares when cold. This candy is softer and creamier than the regular divinity. Approximately forty pieces.

Luci served as hostess at a Christmas party for one hundred fifty needy children one year. A White House staffer, dressed as Santa Claus, distributed gifts to the children, who sat on chairs, sofas, and carpets and opened the gifts with wide-eyed anticipation.

Mrs. Johnson introduced the practice of presenting entertainment at luncheons for Senate wives. Following a spring luncheon for the Senate wives in 1964, a cast headed by Helen Hayes presented scenes from "The White House," a new play by A. E. Hotchner, in the East Room. This was one week before it opened in New York, and the White House introduction was fitting because the production concerned the lives of early Presidents and First Ladies.

At this, as at subsequent luncheons for the wives of Senators, Mrs. Johnson had each of the round tables set with china of a different presidential administration. It was her way of highlighting the history of the mansion, and the china provided a principal topic of conversation.

Mrs. Johnson's dessert recipes include:

STRAWBERRY ICE-BOX PIE

1 cool pastry shell
1 17-ounce package marshmallows
2 tablespoons strawberry juice
1 box frozen strawberries, or
2 cups fresh berries sweetened to taste
1 cup whipping cream

Put marshmallows in double boiler. Add strawberry juice. Cook until marshmallows are melted. Mix strawberries and marshmallows thoroughly. Chill about two hours. Fold in whipped cream to marshmallow mixture and pour into pastry shell. Chill until firm.

ONE-CRUST PASTRY SHELL

1 cup flour
2 tablespoons shortening
1 teaspoon salt
3 tablespoons cold water
1 tablespoon sugar

Sift flour, sugar and salt together. Cut in shortening. Add few drops of water at a time. Roll out and fit in 9-inch pie tin.

BROWNIES

1 cup sugar
2 eggs
3/4 cup flour
1 teaspoon vanilla
1/2 cup melted butter
1/2 cup coarsely chopped nuts
2 squares unsweetened chocolate

Mix all ingredients and spread in greased tin. Bake for 25 minutes and cut in two-inch squares while still warm. Makes sixteen.

A history major at the University of Texas where she also studied journalism, Mrs. Johnson delighted in delving into the history of the White House, and arranging exhibits of its exciting past. At another of her luncheons, she put on display in glass cases in the Blue Room memorabilia associated with past Presidents and First Ladies. There was a pressed glass finger bowl used in the White House by President and Mrs. Lincoln; a cross and earrings mounted with the hair of President and Mrs. James Madison; a letter written by Thomas Jefferson to Benjamin Latrobe, architect for the White House discussing the high cost of stone-hewing for walls around the grounds of the White House, and the engagement book kept by Mrs. Benjamin Harrison.

Adding another historic note to the luncheon, each item on the menu was named for some association with a former occupant of the White House. It included:

Hayes Melon Cup
Veal Abigail Adams
Rice Pilaf First Lady
Asparagus Monticello
Strawberry Sherbet Dolley Madison
Demitasse

Each year the Johnsons invited additional thousands to the White House to breakfasts, buffets, luncheons, briefings, dinners, teas, coffees, receptions, garden parties, late suppers and cookouts.

They were the first Presidential couple to give a cookout at the White House. It was held on the West Terrace (the roof over the swimming pool) outside the State Dining Room, and was in honor of retiring members of Congress. An artificial carpet of grass covered the terrace and grills were set up to broil steaks to the individual taste of each guest. It was not unusual to see the tall, lean President supervising the cooking as the odor of charcoal-broiled steaks wafted on the breeze onto busy Pennsylvania Avenue.

Typically Texan is this Johnson favorite:

NOCHE SPECIALS

8 tortillas	½ lb. sharp cheese, grated
Fat for frying	Jalapenas (hot peppers)

Cut tortillas into quarters and fry in deep hot fat until brown and crisp on both sides. Drain and put about 1 teaspoon grated cheese and slice of jalapeno pepper on each quarter. Place in hot oven until well heated and cheese begins to melt. Serve at once. Makes 32.

President Johnson has done more dancing in the White House than did any of his predecessors. Nearly every evening social event—dinners and receptions especially — ended with dancing and no one twirled around the floor more than the President.

Often after seeing the guest of honor to the door, he returned to the East Room and selected a partner for dancing that continued until midnight or after. Sometimes he stayed with one partner nearly all evening, breaking away for a dance only occasionally with someone else and then going back to his favorite for the night. On other occasions, he tried to give every lady on the guest list a twirl around the floor so she could say, "I danced with the President."

Each summer Presidential scholars from every state gather on the South Lawn to be welcomed by the President, who was once a schoolteacher and still wants to honor "the most precious resource of the United States—the brain power of its young people."

Politics was never forgotten by the Johnsons either. Members of the Democratic National Committee were invited to a number of receptions, and once, three thousand Democratic women, at a campaign conference in Washington, swarmed through the White House at a reception in their honor. Statistically, the biggest "bash" given by the Johnsons, however, was a reception on the South Lawn for nearly seven thousand students. The President also gave a series of briefings for members of Congress, while at the same time Mrs. Johnson entertained the Congressional wives upstairs with a program of invited speakers.

The most spectacular entertainment, in a series of spectaculars, was a "Salute to Congress" in 1965 which was witnessed by very few of the people for whom it was intended. A glittering array of Hollywood and Broadway stars was assembled at the White House to serenade, in music and verse, the Congress which had passed the bulk of the President's legislation for the Great Society. The party was to have been his appreciation gesture. However, the House of Representatives was kept in session until almost midnight wrangling over the Administration's "beautification bill" and its members missed the salute. The show went on to empty seats, regardless.

Mrs. Johnson gave a series of once-a-month "women doers" luncheons to salute women of achievement in various fields of endeavor. These luncheons were held in the second-floor family dining room where a special attraction was the printed wallpaper of scenes of the American Revolution. Often at the same time, President Johnson gave a "stag" luncheon in the State Dining Room.

The menus for the two simultaneous luncheons were different, and they were prepared by different cooks. Zephyr Wright, the Johnsons' long-time family chef from Texas, prepared the luncheons for up to fifty women; and Rene Verdon, the French chef the Johnsons retained from the Kennedy Administration, prepared the President's luncheons until his departure. After that they were prepared by Verdon's successor, Henri Haller, the Swiss-born chef of French parentage, hired by the Johnsons.

Additional recipes from the present Presidential files are:

HAM TURNOVERS

1 pound diced cooked ham
¼ cup chopped onion
1 tablespoon chopped parsley
pinch of thyme
1 rounded tablespoon butter

Melt butter in heavy frying pan and add all other ingredients. Mix with just enough cream sauce to bind mixture together—about ½ cup.

Make a rich pastry dough (enough for a two crust pie). Cut the dough in two inch rounds (after rolling it very thin), and place a heaping teaspoon of the ham mixture on one side of the round. Fold the top over and press edges together with a fork. Brush top with melted butter and bake at 350 degrees until light brown. Makes about 30 turnovers. (Serve with Monroe's Quaking Jelly.

CRANBERRY SALAD

1 envelope plain gelatin
1¼ cups cold water
1 cup sugar
2 cups cranberries
½ cup chopped celery
½ cup chopped nuts
1 teaspoon salt

Cook cranberries in one cup water twenty minutes. Stir in sugar and cook five minutes longer. Soften gelatin in one-quarter cup cold water, add to hot cranberries and stir until dissolved. Cool and when mixture begins to thicken, add celery, nuts and salt. Turn into mold that has been rinsed in cold water. Chill until firm.

WHIPPED SWEET POTATOES WITH TOASTED MARSHMALLOWS

6 medium sweet potatoes boiled until soft
3 tablespoons sugar
3 tablespoons cream
¾ stick butter
1 egg beaten until light
pinch of salt

Peel potatoes and mash. Add other ingredients in order listed. Turn into buttered dish and put in slow oven five to ten minutes. Put marshmallows on top and return to oven until light brown.

SAUCE ESPAGNOLE
(Brown Sauce)

In a heavy saucepan melt one-half cup beef, veal or pork fat—or use clarified butter. Cook in the fat, one small carrot and two medium onions, coarsely chopped, until they just begin to turn golden. Add one-half cup flour and cook, stirring frequently, until the roux takes on a good hazelnut brown color and the carrot and onions are brown. Bring to a boil. Add three cups hot brown stock, one clove garlic, a bouquet garni made by tying together one stalk celery, three or four sprigs parsley, a small bay leaf, and a pinch of thyme. Cook, stirring frequently, until thickened and add three more cups hot stock. Cook the sauce very slowly over very low heat, stirring occasionally, for one to one and one-half hours, or until it is reduced to about three cups. As it cooks, skim off excess fat. Add one-fourth cup tomato sauce or one-half cup tomato puree. Cook the sauce for a few minutes longer. Remove the bouquet garni and strain the sauce through a fine sieve. Add two cups hot stock and cook the sauce slowly for about one hour more, or until it is reduced to about four cups, skimming it from time to time. Cool. Freeze if desired, in small containers.

RED WINE SAUCE

Sauté one tablespoon chopped shallots in one-half tablespoon melted butter. Add one cup dry red wine and one teaspoon white crushed pepper and a bay leaf. Reduce to one-half and add one pint brown sauce. Simmer for about one-half hour and strain through a cheese cloth. This sauce is delicious served with filet of beef.

The dinner given by the Johnsons that most people wanted to attend was in honor of Princess Margaret of England and her husband, Lord Snowdon, the British photographer, on November 17, 1965. It also celebrated President and Mrs. Johnson's thirty-first wedding anniversary.

A choice group of one hundred forty guests were invited to the splendor of the State Dining Room dinner and fifty additional guests for dancing in the East Room. Actor George Hamilton was invited as Lynda Bird's escort that evening and that was the beginning of their courtship.

State dinners given by the Johnsons usually have four courses, soup or fish; a meat entree with vegetables; salad, and a spectacular dessert presented with fanfare, frequently by the Army's Singing Violins. Interestingly, trays for serving tables for ten always carry eleven portions so that no one will have to take the last one.

(*Dinner for King and Queen of Thailand*)
Brook Trout White House
Whole Roast Tenderloin of Beef Bordelaise
Eggplant Nicoise
Cucumber Mousse
Vacherine Chieng Mai

ROAST WHOLE TENDERLOIN OF BEEF BORDELAISE

Trim and tie small whole tenderloin of beef. Use small amount of oil in heavy pan and heat thoroughly. Place tenderloin in pan and brown quickly, turn so that all sides will be browned. Meanwhile, pre-heat oven at 450 degrees. When tenderloin is brown place in oven in same pan used for browning. Bake five minutes, reduce heat to 375 degrees and continue baking for ten to fifteen minutes longer, depending on how well done you want the meat.

Remove the tenderloin from pan and place on rack in warm place where air can get to it. Leave it for fifteen to twenty minutes, transfer to board and slice. Serve immediately with Bordelaise sauce.

BORDELAISE SAUCE

½ cup finely chopped shallots
1 cup red wine
2 cups brown sauce
marrow

Remove marrow from a marrow bone and dice. Poach in salt water for two minutes. Drain and set aside. Add shallots to red wine and cook until wine is reduced to ¼ cup. Stir in brown sauce and boil slowly for 15 minutes. Add the diced marrow, stir and serve.

EGGPLANT NICOISE

1 large eggplant
1½ cup soft bread crumbs
1 tomato
1 onion
1 clove garlic
¼ teaspoon thyme
1 tablespoon finely chopped parsley
1 bay leaf
2 tablespoons salad oil
salt and pepper

Cut eggplant in half and put to boil in cold water until tender (about 30 minutes). Remove from water and set to cool. When cool, remove pulp, leaving the skins unbroken. Set the skins aside. Chop the pulp fine, season with salt and pepper and add bread crumbs. Chop tomato, onion and herbs, keeping all the juice from the tomato. Put the oil in a frying pan and sauté the onion slightly. Add all the other ingredients and let the mixture simmer for a minute or two. Add the eggplant, correct the seasoning and let all fry for five minutes. Remove from heat, cool slightly and fill the shells. Sprinkle the top with bread crumbs, dot liberally with butter and set in the oven to bake at 375 degrees to a nice brown.

CUCUMBER MOUSSE

2 cucumbers
1 tablespoon lemon juice
1 teaspoon worcestershire sauce
¾ teaspoon salt
½ teaspoon white pepper
½ cup mayonnaise
1 envelope gelatin
½ cup heavy cream, whipped stiff

Peel cucumbers, cut in half and remove seeds. Blanch in a cup of boiling water to which lemon juice has been added (about five minutes). Drain well and put through a fine sieve or in blender. When pulp is cool, add worcestershire sauce, salt, pepper and mayonnaise. Soften gelatin with one tablespoon cool water and dissolve in tablespoon of hot water. Add to cucumber mixture. Fold in whipped cream. Turn into a chilled mold or timbale and chill until firm. Garnish with thin sliced cucumbers marinated in french dressing and sprigs of watercress.

VACHERIN CHIENG MAI

Make an 8-egg white dry meringue from your favorite recipe. With a pastry bag, make three 8-inch meringue rings on an oiled and floured baking sheet. Bake the meringues in a very low oven until completely dry. Do not brown. Set aside to cool in a very dry place.

Cut a thin layer of sponge cake in a circle the exact size of the inside of the meringue ring. Place sponge cake on serving plate, and the first meringue ring on top of the cake. Place second meringue on top of the first, "gluing" rings together with whipped cream. Repeat with third ring. Fill center with a marble swirl made by combining one quart vanilla ice cream, one quart strawberry ice cream, and one pint of whipped cream. Season with Kirsch to taste. Place the filled meringue ring in freezer until mixture is firm. Just before serving use pastry tube to flute the edges of the sponge cake with whipped cream. Cover the top of ice cream mixture with fresh strawberries. Glaze with ¼ cup currant jelly which has been melted with two tablespoons water. Granish the rim of the serving plate with strawberries.

For large receptions, Mrs. Johnson likes to serve whole hams and steamship rounds of beef on a buffet decorated with whole cheeses draped with black grapes. Small fancy things to nibble are passed on trays.

Meals in the family dining room are apt to include "fruit of the land" as she terms the jellies, preserves, corn, green beans and other produce from the LBJ Ranch in Texas which are canned or frozen in season for subsequent meals in Washington.

Recipes for vegetable dishes from the Johnson's include:

VEGETABLE BOUQUET

Bibb lettuce leaves
cold cooked asparagus, whole spears
cold braised celery, four to five stalks
cold whole string beans
hearts of palm (fine slices)
dressing (see recipe following)
garnish: sliced pitted black olives
small strips of pimento
grated hard-boiled eggs

Bind green beans and hearts of palm together with dressing. Make a mound in the center of the platter. Arrange lettuce around mound. Arrange asparagus and celery in alternating mounds all around the beans. Pour dressing over all and garnish liberally with olives, pimento and grated eggs.

DRESSING (For Vegetable Bouquet)

4 chopped shallots
1 finely chopped garlic clove
1 pinch white pepper
1 lemon (use juice only)
½ teaspoon English mustard
2 egg yolks
½ juice glass of red wine vinegar
1½ cup of pure olive oil
salt to taste
½ teaspoon of Accent

Put shallots, garlic, peppers, lemon juice, English mustard, egg yolk and vinegar in stainless steel bowl. Mix thoroughly with a steel whip. Blend olive oil slowly into mixture.

SPINACH SOUFFLÉ

¼ cup chopped onion
2 tablespoons butter
2 tablespoons flour
1 cup whole milk or light cream
½ teaspoon salt
⅛ teaspoon pepper
3 eggs, separated
1 cup chopped cooked spinach
½ cup grated cheese

Sauté onion in small amount of butter and set aside. Melt butter, add flour and then milk, gradually; season. Add the sauted onion. Beat yolks until thick and lemon-colored. Stir into white sauce; add spinach and cheese. Fold in stiffly beaten egg whites and turn into greased casserole. Set in pan of hot water and bake in moderate oven (350°) for about fifty minutes. Serve at once. Serves four.

Hot freshly-baked breads are served at all the mansion's dining tables: in the State Dining Room, the family dining room, the staff cafeteria, and the Navy "mess" for top level officials in the White House.

Bess Abell, who has a leading hand in making up menus for state dinners (in consultation with Mary Kaltman, coordinator of food, the chef, and representatives of the State Department's Protocol Office), delights in naming dishes after the guests of honor. It goes over well with the guests, also.

When the Johnsons gave a black tie dinner in honor of three brides-to-be in his official family: Donna Erickson, who married Vice President Humphrey's son, Robert; Margaret McNamara, daughter of Secretary of Defense Robert McNamara, who married Barry Carter; and Margaret Hickman, who married Richard Wirtz, son of the President's Secretary of Labor Willard Wirtz, Bess dreamed up a romantic menu and decorations.

At each table was a black and white cartoon of an historic romance such as Romeo and Juliet's; and Cupid Glace (raspberry ice with fresh peaches) crowned the menu.

Of all the thousands of guests invited to dinner at the White House by the Johnsons, only one has declined to eat the Presidential fare. Occasionally a guest will note in his acceptance of an invitation that he doesn't eat meat or has a dietary problem and the White House serves him a fruit platter rather than what is on the menu. The one guest who wouldn't touch White House food was King Faisal of Saudi Arabia, who had a stomach problem, and ate only chopped broiled lamb and yogurt prepared especially for him.

When he came to Washington for a state visit he brought along his royal chef, who prepared his food in the kitchen at the President's Guest House, across Pennsylvania Avenue from the White House. The chef arrived at the White House just before the state dinner wearing the traditional turban and baggy silk trousers. He brought with him four black briefcases which were set up on a table just behind the King's seat at the table.

As each course of White House fare was served other guests at the table, the King's royal waiter, also in turban and flowing trousers, removed a plate from a briefcase and set it before His Majesty.

It was a slightly different procedure from the occasion when the King's older brother was honored at a White House dinner given by a previous President. That time, the King ate Presidential fare, but one of his bodyguards stood at his shoulder, his hand on the dagger in his belt, and his eyes watching suspiciously the White House butler who served his royal master.

The Johnsons invite an official party of ten from the country of each chief of state honored at a dinner and allow the visiting chief to choose who shall accompany him. It may be anyone from his shoeshiner to his Secretary of State. Thus, they avoid a problem that confronted President Kennedy when a chief of state brought his "girl friend" instead of his wife to Washington and wanted her invited to the dinner in his honor.

The White House replied that because of space limitations, only visiting officials of cabinet rank could be invited, whereupon the state visitor had his "girl friend" sworn in as Secretary of State for the evening, and in that capacity, she got to attend the State Dinner.

Accidents happen at White House dinners the same as everywhere else, and one night it seemed that one gentleman guest of the Johnsons had more than his share. During the dinner he was talking animatedly with his attractive dinner partner, who wore a new white silk gown. As he waved his hand in a conversational gesture, he accidentally hit her wine glass lifted to her lips and splashed wine over the front of her gown as well as the table. As if this were not enough, a few moments later, he leaned back in his chair, putting his weight on the two back legs, and it crashed to the floor. A butler silently moved in with another chair for the guest, swept up the splinters and removed them as unobtrusively as possible. Not an hour later, the guest was seated in the East Room watching the after-dinner entertainment when he leaned back again in one of the dainty gold chairs. It, too, crashed to the floor. With great equanimity, the gentleman moved to another chair and finished his evening with apparent enjoyment.

Rated the most exciting and emotional of all the social events held during the Johnsons' first four years in the mansion was the wedding of their nineteen-year-old daughter, Luci Baines.

She was married to Patrick John Nugent of Waukegan, Illinois, on August 6, 1966, in a sentimental ceremony at Washington's National Shrine of the Immaculate Conception, the largest Catholic Church in the United States. A reception followed in the White House.

Plans for the wedding, first of a presidential daughter in three decades, were begun six months ahead and the coveted invitations to seven hundred selected "personal friends and relatives" were mailed early in July.

Before the wedding at twelve o'clock noon on that hot summer Saturday, some members of the White House staff worked all night decorating the Shrine for the ceremony, and the executive mansion for the reception which followed. Used throughout the Shrine were willowy greenery—locust and ficus trees—decorated with white roses, lily-of-the-valley, and babies breath.

At the White House, a white canopy was placed over the Jacqueline Kennedy Garden, where reception guests assembled before going to the Blue Room to greet the bridal couple and their parents. The tent had a pink lining and the poles and ropes were decorated with pink carnations and greens.

The long corridor through which the guests walked

to the Blue Room was lined with pots of white petunias and pink geraniums. The gold chandelier in the State Dining Room was used as a holder from which greens and flowers cascaded; and compotes of yellow, white, blue and orange flowers decorated the table.

Swags of greens highlighted with flowers festooned the four columns in the entrance hall on the State floor, and gold *torchères* in the hall were decorated with greens and flowers. Red roses and rubrum lilies, accented by pink and white flowers, highlighted the historic Red Room; green Nicotiana, lavender and pink and white flowers decorated the Green Room. Arrangements of white, yellow, red and blue flowers decorated the East Room. Here, towering toward the ceiling, was the fourteen-tiered wedding cake baked by Ferdinand Louvat, French pastry chef at the White House. There were seven layers of the summer fruit cake (recipe below) with a tier of sugar arches framing sugar lilies-of-the-valley between each layer of cake. On top was a chocolate cake—the bride's cake, which the couple took on their honeymoon. And on top, in lieu of the traditional miniature bride and groom, were real lilies-of-the-valley.

LUCI'S WEDDING CAKE

1¾ cups sifted cake flour
1 teaspoon double-acting baking powder
¼ teaspoon salt
½ cup butter
¾ cup sugar
5 egg whites, unbeaten
¾ cup chopped candied pineapple
1 cup chopped pecans
½ cup soaked raisins
½ teaspoon almond extract
½ teaspoon vanilla extract

Cover ½ cup white seedless raisins with apple juice and let soak in refrigerator two or three days (or until raisins are plump). Drain and spread on towel to dry surface moisture.

Sift flour once and measure. Add baking powder and salt and sift together three times. Cream butter thoroughly and add sugar gradually, creaming together until light and fluffy. Add egg whites, one at a time, beating thoroughly after each. Add fruit, nuts and flavoring and mix well. Add flour, a little at a time, beating after each addition until smooth. Bake in a loaf pan which has been greased, lined with heavy paper and greased again. Bake in a slow oven (300°) about one hour and fifteen minutes or until done. For loaf cake, use pan 8 x 4 x 3 inches which serves eight to ten persons. Frost as desired.

It was a romantic dream wedding and reception and before the newlyweds left for their honeymoon, Luci looked down from the second floor balcony to the guests gathered on the lawn and said goodbye to each of her attendants calling them by name. And, she added, "I want to say goodbye to everybody else. It has been a beautiful, wonderful day. The most magnificent day of our lives."

Each new social event held by the Johnsons seemed to outshine the preceding one, with added details of interest that made it especially appropriate for a particular occasion or a guest of honor.

Social Secretary Bess Abell was always on the lookout for new ideas for entertainment, and the White House parties she planned were composites of many things—the things she had seen or heard somewhere else and adapted for the situation in the White House.

Ideas came from people everywhere—members of Congress, people in show business, talent agents who wanted to promote their proteges, and friends who were ever alert to her needs.

In the same way that she adapted these ideas for use in Presidential entertaining, you may adapt the ideas presented in this book to your own situation and thereby establish a reputation as a creative hostess whose parties have a new dimension. And have fun yourself, while you do it.

The Marine Band provides music for a buffet supper on the South Lawn. The food was served under the red and white striped tent in the background.

MENUS SERVED BY THE JOHNSONS IN THE WHITE HOUSE

SPRING LUNCHEON MENUS

(*Ladies luncheon in honor of Mrs. Lester Pearson, wife of the Canadian Prime Minister*)

Jellied Consommé
Pheasant Supreme
Wild Rice
String Beans Amandine
Broiled Tomatoes
Strawberry Mousse
Demitasse
Wines: Beaulieu Beaumont Pinot Noir
Paul Ruinart Blanc de Blanc 1953

(*Ladies luncheon for Women-Doers*)

Vichyssoise
Filet of Beef
Broiled Tomatoes
Asparagus Hollandaise
Strawberries White House
Wine: Tavel Rosé 1961

(*For Senate Wives*)

Prosciutto and Melon
Supreme of Chicken Maryland
Purée Favorite
Fresh Fruit au Cointreau
Demitasse

(*Honoring Willy Brandt, Governing Mayor of Berlin*)

Melon-balls in Port Wine
Leg of Lamb White House
Purée Favorite
Braised Endive
Chocolate Mousse
Demitasse
Wines: Bernkasteler Schlossberg Spatlese Doktor Thanisch 1961
Inglenook Cabernet Sauvignon
Piper-Heidsieck 1959

SUMMER LUNCHEON MENUS

(*Honoring the Shah of Iran and Empress Farah*)

Eggs Florentine

Broiled Lobster
Lattice Potatoes Puree Favorite
Strawberries Farah
Demitasse
Wines: Chassagne Montrachet 1961
Paul Ruinart Blanc de Blanc 1955

(*For Senate Wives*)
Monroe's Tomato Jelly served with Ham Turnovers
Jefferson's Filled Pannequaiques
Adams' Fourth of July Peas
Vegetable Bouquet
Creme Brulée Hermitage

(*One in a series for Women-Doers*)
Cream of Mushroom Soup
Avocado with Crabmeat
Fresh Asparagus
Strawberry Mousse
Demitasse
Wine: Inglenook Pinot Chardonnay

(*Honoring Mrs. Inonu, wife of the Prime Minister of Turkey*)
Consommé Madriléne
Broiled Tomatoes
Crepe Suzettes Mevhibe
Asparagus Hollandaise
Green Salad Macedoine of Fruit
Wine: Almaden Pinot Chardonnay

(*Honoring the Prime Minister of Turkey, Ismet Inonu*)
Melon au Porto
Breast of Chicken Georgina
Rice Pilaf Zucchini
Glace Inonu
Demitasse
Wine: Almaden Grenache Rosé
Dry Monople 1955

(*Honoring the Prime Minister of Australia, Sir Robert Gordon Menzies*)
Cold Cream of Senegalese
Veal au Marsala
Rice à la Grecque
Green Beans with Almonds
Macedoine of Fruits

Demitasse
Wine: Inglenook Cabernet Sauvignon

FALL AND WINTER LUNCHEON MENUS

(*Honoring West German Chancelor Ludwig Erhardt served at the LBJ Ranch in Texas*)

Savory Cheese Custard
Trout Amandine   Parsley Potatoes
Asparagus with Lemon Sauce
German Chocolate Cake
Demitasse

(*Stag luncheon for the Executive Council, AFL-CIO*)

Onion Soup
Filet Mignon Blanchette
Petits Pois à la Francaise
Chateau Potatoes
Almond Supreme
Demitasse
Wine: Beaulieu Beaumont Pinot Noir

(*Honoring Canadian Prime Minister Lester Pearson*)

Lobster Thermidor
Filet Mignon Choran
O'Brien Potatoes   Asparagus Polonaise
Strawberries Romanoff
Demitasse
Wines: Beaulieu Beaumont Pinot Noir
Paul Ruinart, Blanc de Blanc 1953

Filet de Pompano Amandine
Rack of Lamb Vert Pre
Vegetables Nicoise
Watercress and Endive Salad
Brie Cheese
Charlotte Royale
Demitasse
Wines: Almaden Pinot Blanc
Chateau Haut-Brion 1959
Moët and Chandon 1959

(*One in a series for Women-Doers*)

Jellied Consommé
Pheasant Virginia

Asparagus Broiled Tomatoes
Popovers
Strawberry Meringue
Wine: Pinot Chardonnay

(*For ladies only*)

Filet Mignon White House
Fresh Asparagus
Broiled Tomatoes Garden Salad
Lemon Supreme

SPRING DINNER MENUS

(*Honoring His Majesty, Hussein I, King of the Hashemite Kingdom of Jordan*)

Vol au vent of Seafood Newberg
Filet of Beef Nemours
Creamed Spinach
Endive and Watercress Salad
Brie Cheese
Strawberries Romanoff
Wines: Wente Pinot Chardonnay
Pommard Billardet Hospices de Beaune 1959
Paul Ruinart Blanc de Blanc 1959

(*Stag Dinner for Business Leaders*)

Brook Trout Grenobloise
Roast Sirloin Beaucaire
String Beans with Almonds
Salad Mimosa
Brie Cheese
Peach Melba
Wines: Almaden Mountain White Chablis
Beaune, 1959
Louis Latour Dry Monopol 1959

(*Stag Dinner for Labor Leaders*)

Crèpes of Crabmeat Nantua
Chicken Chasseur
Wild Rice
String Beans
Asparagus Vinaigrette
Strawberry Mousse
Petits Fours
Wines: Wente Pinot Chardonnay
Laurent Perrier
Blanc de Blanc 1959

(*Honoring the President of Ireland, Eamon de Valera*)

Baby Lobster en Bellevue
Roast Sirloin President
French Peas
Dauphine Potatoes
Green Salad
Brie Cheese
Baked Alaska

Wines: Krug Chablis
Pommard, Billardet Hospices de Beaune 1959
Dom Perignon 1959

(*Honoring the Prime Minister of Israel and Mrs. Eshkol*)

Eggs Tarragon
Rack of Lamb Vert Pre
Green Beans aux Champignons
Watercress and Endive Salad
Strawberry Sherbet

Wines: Almaden Pinot Chardonnay
Chateau Gruaud-Larose 1959
Piper-Heidsieck 1959

(*Honoring the Prime Minister of Denmark and Mrs. Krag*)

Crabmeat Ravigote
Tournados Baltimore
Wisconsin Potatoes
Watercress and Endive Salad
Brie Cheese
Charlotte Russe

Wines: Krug Chablis
Pommard Billardet Hospices de Beaune 1955
Piper-Heidsieck 1959

(*Honoring West German Chancellor Ludwig Erhard*)

Vichyssoise
Lobster Thermidor
Berny Potatoes
Creamed Spinach
Green Salad
Mousse of Roquefort
Strawberries Romanoff

Wines: Inglenook Pinot Chardonnay
Dom Perignon 1955

(*Honoring the Prime Minister of Great Britain and Mrs. Harold Wilson*)

Crabmeat Chesapeake
Roast Sirloin of Beef
Artichokes White House
New Potatoes
Green Salad
Ham Circles
Strawberry Creme York

(*Honoring the President of Costa Rica and Mrs. Orlich*)

Cream Baltimore
Filet Mignon Orlich
Creamed Spinach
Green Salad
Mousse of Roquefort
Coffee Parfait

Wines: Almaden Golden Sherry
Pommard Billardet Hospices de Beaune 1959
Piper-Heidsieck 1959

FALL AND WINTER DINNER MENUS

(*Stag Dinner for the Business Council*)

Filet de Sole Duglere
Roast Sirloin Nemours
Potatoes Lorette
Creamed Spinach
Green Salad
Brie Cheese
Peaches Cardinal

Wines: Inglenook Pinot Chardonnay
Clos de la Roche 1955
Piper-Heidsieck 1955

(*Honoring the Vice President, the Chief Justice and the Speaker of the House of Representatives*)

Sole Nina (named for Mrs. Earl Warren, wife of the Chief Justice)
Pheasant Muriel (named for Mrs. Hubert H. Humphrey, wife of the Vice President)
Wild Rice Croquettes
Asparagus
Winter Garden Salad with Minnesota Cheese
Harriet Souffle (named for Mrs. John McCormack, wife of the Speaker)

(*Honoring the President of the Republic of Italy and Mrs. Segni*)

Delices of Maryland
Filet of Beef Wellington
Waffle Potatoes
String Beans Amandine
Endive and Watercress Salad
Brie Cheese
Bavarois Ceylan

(*For United States Senators and Their Wives*)

Filet of Sole Murat
Breast of Capon Valley D'Auge
Rice Pilaf
Èpinards en Branches
Chocolate Mousse
Wine: Inglenook Pinot Chardonnay

* * *

Gnocchi with Cheese Sauce
Broiled Breast of Chicken Americaine
Purée Californienne
Salad Mimosa
Colonial Rum Cake
Wine: Inglenook Pinot Chardonnay

(*For United States Senators*)

Cream of Mushroom Soup
Breast of Chicken Washington
French Peas
Green Salad
Chocolate Mousse
Wine: Almaden Pinot Blanc

(*Honoring the Prime Minister of Great Britain, Sir Alec and Lady Douglas-Home*)

Shrimp Mousse White House
Filet Mignon Choisy
Creamed Spinach
Salad Mimosa
Brie Cheese
Dame Blanche Diplomat
Wines: Wente Pinot Chardonnay
Musigny Jacques Prieur 1959
Dom Perignon 1959

(Celebrating President and Mrs. Johnson's Thirtieth Wedding Anniversary)

Melon of the Gods and Prosciutto
Presidential Filet of Beef
with Broiled Mushrooms
Kumquats
Spinach Soufflé
Mixed Green Salad
Anniversary Cake Circa 1934
Wines: Clos de la Vigne au Saint 1959
Dry Monopole 1959

(Honoring Her Royal Highness, Princess Irene of Greece)

Crabmeat Crêpes Irene
Tournédos
Saffron Rice Pilaf
Fresh Spinach Ring
Endive and Watercress
Brie Cheese
Strawberries Monticello

(Honoring England's Princess Margaret and Lord Snowdon and in celebration of President and Mrs. Johnson's Thirty-first Wedding Anniversary)

Pompano Amandine
Boned Roasted Squab stuffed with Wild Rice
Artichoke Bottoms filled with Pureed
Peas and Carrots
Green Salad
Brie Cheese
Praline Glace
(A concoction of coffee ice cream, pralines, and whipped cream)

A table set for a state dinner given by President and Mrs. Lyndon B. Johnson.

Tables set for a state dinner in the Rose Garden at the White House.

BIBLIOGRAPHY

Adams, Abigail, *Letters of Mrs. Adams, the Wife of John Adams,* Boston: Wilkins, Carter & Co., 1848.

........................, *News Letters of Abigail Adams, 1788-1801,* edited by Stewart Mitchell, Boston: Houghton Mifflin Co., 1947.

Adams, Charles Francis, ed., *Memoirs of John Quincy Adams.* Philadelphia: J. B. Lippincott Co., 1874-1877, Vols. IV-VIII.

Adams, Rachel, *On the Other Hand.* New York, Evanston, and London: Harper & Row, 1963.

Aikman, Lonnelle, *The Living White House.* Washington, D. C.: The White House Historical Association. 1966.

Alexander, Holmes, *The American Talleyrand.* New York and London: Harper & Brothers, 1935.

Ames, Mary Clemmer, *Ten Years in Washington.* Hartford, Conn.: A. D. Worthington & Co., 1873.

Anderson, Isabel W., *Presidents and Pies.* Boston and New York: Houghton Mifflin Co., 1920.

......................., "In and Out of the White House," *Nations Magazine,* February 1928.

Anderson, Mary Frances, "The Costumes of the Mistresses of The White House," *Americana,* New York, 1929.

......................, "The White House China," Lenox, Inc., Trenton, New Jersey, April 1937.

Bache, Rene, "Entertaining At The White House," *The Saturday Evening Post,* November 1904.

Banks, Elizabeth L., "Easter Egg Rolling in Washington," *The Quiver,* London, April 1, 1899.

......................, "Public Receptions at The White House," *Cassell's Magazine,* July 1898.

......................, "Sunday at The White House," *The Quiver,* July 1898.

Beale, Harriet S. Blaine, ed., *Letters of Mrs. James G. Blaine.* New York: Duffield & Co., 1898.

Beale, Marie, *Decatur House and Its Inhabitants.* Washington: National Trust for Historic Preservation, 1954.

Beard, Alice and Frances Rogers, "The First First Lady in The First White House," *Independent Woman,* January 1945.

Blair, Emily Newell, "If You Were Gadding About Washington," *Good Housekeeping,* March 1934.

Bobbe, Dorothie, *Abigail Adams, the Second First Lady.* New York: Minton Balch & Co., 1929.

......................, *Mr. and Mrs. John Quincy Adams.* New York: Minton, Balch & Co., 1930.

Booth, Alice, "America's Twelve Greatest Women, Grace Coolidge," *Good Housekeeping,* July 1931.

Braxton, Fairfax, "Wives of the Presidents," *National Republic,* September, 1928.

Brent, Portia, "The Social Side, Jefferson's Birthday," *The New Age,* 1910.

Briggs, Emily E., *The Olivia Letters,* New York and Washington: The Neale Publishing Co., 1906.

Buchanan, Jr., Wiley T. with Arthur Gordon, *Red Carpet at The White House.* New York: E. P. Dutton & Co., Inc. 1964.

Burke, Pauline Wilcox, *Emily Donelson of Tennessee.* Richmond, Va.: Garrett & Massie, Inc., 1941.

Butler, Joseph G., *Life of William McKinley and History of National McKinley Birthplace Memorial.* Cleveland, 1924.

Butt, Archibald W., *The Letters of Archie Butt, edited by Lawrence F. Abbott,* Garden City, N.Y.: Doubleday, Page & Co., 1924.

......................, *Taft and Roosevelt, the Intimate Letters of Archie Butt.* Garden City, N.Y.: Doubleday, Doran & Co., Inc., 1930.

Cantacuzene, Princess, "Society and Statesmen," *The Saturday Evening Post,* August 5, 1922.

Carvallo, Emma De, "Dining With the President," *The Ledger Monthly,* April 1902.

Cavanah, Frances, *Children of The White House.* New York and Chicago: Rand McNally & Co., 1936.

Chamberlain, Ivory, *Biography of Millard Fillmore.* Buffalo: Thomas & Lathrops, 1856.

Chase, Jefferson, "Your Host in The White House," *Vanity Fair,* March 1931.

Chassler, Sey, "Pageant in The White House," *Pageant,* December 1950.

Clark, Allen C., *Life and Letters of Dolly Madison.* Washington: W. F. Roberts Co., 1914.

Coe, Charles H., "Miss Alice Roosevelt and The White House," *The Ledger Monthly,* April 1902.

Coleman, Elizabeth Tyler, *Priscilla Cooper Tyler and the American Scene.* University of Alabama Press, 1955.

Coleman, Edna M., *Seventy-five Years of White House Gossip; From Washington to Lincoln.* Garden City, N. Y.: Doubleday, Page & Co., 1926.

......................., *White House Gossip, From Andrew Johnson to Calvin Coolidge.* Garden City, N. Y.: Doubleday, Page & Co., 1927.

Colquitt, Dolores Bioteuillet, "The Republican Court," *D.A.R. Magazine,* February 1923.

Coolidge, Calvin, *The Autobiography of Calvin Coolidge.* New York: Cosmopolitan Book Corp., 1929.

Coolidge, Mrs. Calvin, "Our Family Pets," *American Magazine,* December 1929.

Coolidge, Mrs. Calvin, "Home Again," *American Magazine,* January 30, 1930.

Coolidge, Mrs. Calvin, "When I Became the First Lady," *Americana,* September 1929.

Coolidge, Mrs. Calvin, "Making Ourselves at Home in The White House," *American Magazine,* November, 1929.

Coolidge, Grace A. "How I Spent My Days at The White House," *American Magazine,* October 1929.

Cresson, William Penn, *James Monroe.* Chapel Hill: The University of North Carolina Press, 1946.

Criss, Mildred, *Abigail Adams, Leading Lady.* New York: Dodd, Mead & Co., 1952.

Crockett, David, *The Life of Martin Van Buren.* Philadelphia: R. Wright, 1835.

Crook, William H., *Memories of the White House, edited by Henry Rood.* Boston: Little, Brown & Co., 1911.

Culbertson, Mary, "Our Hospitable and Home-Loving First Family," *Better Homes and Gardens,* May, 1934.

Curtis, Charlotte, *First Lady.* New York: Pyramid Publications, 1962.

Cutts, Lucia B., ed. *Memoirs and Letters of Dolly Madison.* Boston and New York: Houghton Mifflin Co., 1886.

Dall, Anna Roosevelt, "When The White House Gives a Party," *Cosmopolitan,* March 1935.

Daniels, Jonathan, "The Lady from Independence," *McCall's* April 1949.

Dean, Elizabeth Lippincott, *Dolly Madison, The Nation's Hostess.* Boston: Lothrop, Lee and Shepard Co., 1928.

Deman, Ruth Van, "U.S. Kitchen No. 1," *Journal of Home Economics,* February 1936.

Dyer, Brainerd, *Zachary Taylor.* Baton Rouge, La.: Louisiana State University Press, 1946.

Earle, Alice Morse, "Presidential China," *China Collecting in America,* New York, 1924.

Ellet, Mrs. E. F., *Court Circles of the Republic.* Philadelphia: Philadelphia Publishing Company, 1872.

Ervin, Janet Halliday, *The White House Cookbook.* Chicago: Follett Publishing Company, 1964.

Essary, Helen, "The President's Boss . . . Bess Truman," *Look,* March, 1949.

Essary, Helen, "Upper Crust," *The Senator,* February 18, 1939.

Faulkner, Violet, "White House Recipes," *The Sunday Star Magazine,* January 20, 1957.

Fawcett, Waldon, "White House Receptions," *The Brown Book of Boston,* May 1901.

......................, "Easter Egg Rolling on The White House Lawn," *The Woman's Home Companion,* April 1901.

Fields, Alonzo, "Looking Over the President's Shoulder," *The Ladies Home Journal,* May, June, July, August 1960.

Ford, Elizabeth, "Lincoln's Summer White House at Soldier's Home," *American Motorist,* September 1932.

Fox, Frances Margaret, "The First Baby of The White House," *The Christian Register,* July 30, 1925.

......................, "On the Trail of the Easter Egg," *The Christian Register,* April 9, 1925.

......................, "The Easter Monday Egg Rolling at Washington," *American Motorist,* April 1931.

Fremont, Jessie Benton, *Souvenirs of My Time.* Boston: D. Lothrop & Co., 1887.

Furman, Bess, *White House Profile.* Indianapolis and New York: The Bobbs-Merrill Company, Inc., 1951.

......................, "The Oval Room: Eisenhoweriana," *The New York Times Magazine,* November 30, 1948.

Furnas, Helen, "She Runs The White House," *Reader's Digest,* April 1944.

Gelb, Arthur and Barbara, "Culture Makes a Hit at The White House," *The New York Times Magazine,* January 28, 1962.

Gerry, Margarita Spalding, "Rutherford B. Hayes in The White House—Reminiscences of William H. Crook," *Century Magazine,* March 1909.

......................, "Andrew Johnson in the White House—Reminiscences of William H. Crook," *Century Magazine,* September and October 1908.

Gilfond, Duff, "Alice Roosevelt Longworth," *Plain Talk,* June 1930.

Goebel, Dorothy Burne and Julius Goebel, Jr., *Generals in The White House.* Garden City, N. Y.: Doubleday, Doran & Co., Inc., 1945.

Chautauquan, August 1899.

Goodwin, Katharine Calvert, "White House Invitations," *D.A.R. Magazine,* June 1920.

Gouverneur, Marian, *As I Remember.* New York and London: D. Appleton & Co., 1911.

Grant, Jesse R., *In the Days of My Father, General Grant.* New York and London: Harper & Brothers, 1925.

Graves, Katherine Busbey, "Mrs. Taft's Home Making," *Good Housekeeping,* September 1911.

Griffis, William E., *Millard Fillmore.* Ithaca, N.Y.: Andrus & Church, 1915.

Grimsley, Elizabeth Todd, "Six Months in The White House—1861," *Illinois State Historical Society Journal,* October 1926-January 1927.

Hackett, Catherine I., "Washington's 'Season' Opens," *The Woman Citizen,* January 1927.

Halle, Rita S., "That First Lady of Ours," *Good Housekeeping,* December 1933.

Hard, Anne, "The First Lady of the Land," *Pictorial Review,* September 1926.

Harrison, Sarah, *The House Keeper's Pocket-Book.* London, 1775.

Harcourt, Helen, "The Early Days of The White House," *Americana,* March 1911.

......................., "The President's New Year Receptions, Then and Now," *Americana,* January 1911.

Hay, John, "Life in the White House in the Time of Lincoln," *Century Magazine,* November 1890.

Helm, Edith B., *The Captains and the Kings.* New York: G. P. Putnam's Sons, 1954.

Homlock, Charles and Norris, Margaret, "Flowers For First Ladies," *The Saturday Evening Post,* November 28, 1931.

Hirst, Francis W., *Life and Letters of Thomas Jefferson.* New York: Macmillan Co., 1926.

Hobart, Jennie, *Second Lady.* New York: privately printed (Mt. Vernon, N.Y.: W. E. Rudge), 1933. Material included in the main portions of the book appeared under the title "Second Lady" in *The Saturday Evening Post,* June 29, 1929.

Hogan, E. Pendleton, "Christmas at The White House," *American Motorist,* December 1932.

Holliday, Carl, "Lincoln's First Levee," *Illinois State Historical Society Journal,* October 1918.

Hoover, Irwin H., *Forty-two Years in The White House.* Boston and New York: Houghton Mifflin Co., 1934. Also *The Saturday Evening Post* series of articles, edited by Wesley Stout, February-May 1934; August-November, 1934.

Howard, Clifford, "Egg Rolling in The President's Big Yard," *The Ladies Home Journal,* April 1899.

Howard, Oliver Otis, *General Taylor.* New York: D. Appleton & Co., 1892.

Howe, George Frederick, *Chester A. Arthur, a Quarter-century of Machine Politics.* New York: Dodd, Mead & Co. 1934.

Hunt, Gaillard, "Mrs. Madison's First Drawing-Room," *Harper's Monthly Magazine,* June 1910.

Hunt, Gaillard, *The Life of James Madison.* New York: Doubleday, Page & Co., 1902.

Hurd, Charles, *Washington Cavalcade.* New York: E. P. Dutton & Co. 1948.

........................, *The White House, a Biography.* New York and London: Harper & Brothers, 1940.

Hurst, Fannie, "The First First Lady," *The Democratic Digest,* March 1935.

Jaffrey, Elizabeth, *Secrets of the White House.* New York: Cosmopolitan Book Corp., 1927.

James, Marquis, *The Life of Andrew Jackson.* Indianapolis: The Bobbs-Merrill Company, 1938.

Jeffries, Ona Griffin, *In and Out of The White House.* New York: Wilfred Funk, Inc., 1960.

Jennings, Paul, *A Colored Man's Reminiscences of James Madison.* Brooklyn: George C. Beadle, 1865.

Jensen, Amy LaFollette, *The White House and Its Thirty-Four Families.* New York, Toronto and London: McGraw-Hill Book Company, 1958, 1962, 1965.

Johnson, Monroe, "Mrs. Monroe: Her Doctrine," *National Republic,* December 1930.

Keckley, Elizabeth Hobbs, *Behind the Scenes.* New York: G. W. Carleton & Co., 1868.

Kimball, Marie, *Thomas Jefferson's Cook Book.* Richmond: Garrett & Massie, 1938.

........................, "Epicure of The White House: Thomas Jefferson," *Virginia Quarterly Review,* January 1933.

Kinnaird, Virginia, "Mrs. Lincoln as a White House Hostess," *Illinois State Historical Society Paper for 1938.*

Klapthor, Margaret Brown and others, *The First Ladies Cook Book.* Parents Magazine Press. 1965.

Knox, Paul, "White House Reception," *Reader's Digest,* May, 1936.

Kopp, Clara B., "Washington's New Year," *National Republic,* January 1929.

Kutner, Nanette, "My Life in The White House—Robert Taft," *The American Weekly,* March 23, 1952.

Langford, Laura Holloway, *The Ladies of The White House.* Philadelphia: Bradley & Co., 1881.

Leech, Margaret, *Reveille in Washington.* New York and London: Harper & Brothers, 1941.

Lester, Barnett, "What The Roosevelts Like in Swing Music," *The Senator,* June 17, 1939.

Lewis, Alice Ewing, "In The White House," *Munsey's Magazine,* September, 1896.

Lillie, Lucy C., "The Mistress of The White House," *Lippincott's Magazine,* July 1887.

Lincoln, Abraham, *Abraham Lincoln's Autobiography* with additional biographical material by Robert Dale Richardson, Boston: Beacon Press, 1947.

Lincoln, Evelyn, *My Twelve Years With John F. Kennedy.* New York: David McKay Co., Inc., 1965.

Lincoln, Natalie Sumner, "Ancestry of the Wives of the Presidents of the United States," *D.A.R. Magazine,* April-May 1924.

Lisitzky, Gene, *Thomas Jefferson.* New York: The Viking Press, 1933.

Logan, Mrs. John A., "The Gracious Lady of the White House," *Frank Leslie's Popular Monthly,* July 1899.

Logan, Mary S., *Thirty Years in Washington.* Hartford, Conn.: A. D. Worthington & Co., 1901.

Long, Ava, "A Three a Day in The White House," *The Ladies Home Journal,* December, 1935.

......................., "Company Coming," *The Ladies Home Journal,* October, 1933.

......................., "900,000 Callers a Year," *American Magazine,* June, 1933.

Longworth, Alice Roosevelt, "Some Reminiscences," *The Ladies Home Journal,* March 1933.

Longworth, Alice Roosevelt, *Crowded Hours.* New York and London: Charles Scribner's Sons, 1933.

Mabie, Janet, "Entertaining at The White House," *The Pictorial Review,* October, 1932.

Maclay, Edgar Stanton, "The Social Side of Washington's Administration," *D.A.R. Magazine,* April 1918.

Mansur, Caroline E., *The Virginia Hostess.* Mount Vernon, Va., 1960.

Marshall, Ann Parks, *Martha Washington's Rules for Cooking.* Washington, D. C.: Ransdell, Inc., 1931.

Martin, Edward Winslow (pseudonym of James Dabney McCabe), *Behind the Scenes in Washington.* New York: Continental Publishing Co., 1873.

Martin, Oliver, "America's Best Known Residence and Some of Its Neighbors," *The Transmitter,* December 1933.

McAdoo, Eleanor Randolph, and Gaffey, Margaret Y., *The Woodrow Wilsons,* New York: The Macmillan Co., 1937.

McGregor, "Social Activities of The White House," *Harper's Weekly,* April 25, 1914.

Miller, Jennes, Magazine, "New Year's Customs in Washington," January 1890.

Murdock, Henrietta, "At Home in The White House," *The Ladies Home Journal,* November, 1944.

......................., "First Ladies on Parade," *The Ladies Home Journal,* November, 1944.

Murray, John A., "Simply Superb," *National Historical Magazine,* 1939, LXXLLL, No. 1.

Nelson, Anson, and Fanny Nelson, Memorials of *Sarah Childress Polk.* New York: A.D.F. Randolph & Co., 1892.

Nelson, Henry Loomis, "White House Receptions," *Harper's Weekly,* July 2, 1892.

Nesbitt, Henrietta, *White House Diary,* Garden City: New York: Doubleday & Company, 1948.

Nevins, Allan, ed., *Letters of Grover Cleveland, 1850-1908.* Boston and New York: Houghton Mifflin Co., 1933.

Nevins, Allan, ed., *Polk, the Diary of a President, 1845-1849.* London and New York: Longmans, Green & Co., 1929.

Nicholls, Joanna R., "Ninety-Seven Years in The White House," *Frank Leslie's Popular Monthly, March 1897.*

O'Conor-Eccle, C., "President and Mrs. McKinley at The White House," *Woman at Home,* London, October, 1897.

O'Malley, Frank Ward, "The War Whirl in The White House," *The Century Magazine,* May 1918.

O'Neil, Jean C., "How the President Dines," *The Search-Light,* April 1911.

Orr, Lyndon, "The Most Brilliant Days of Old-Time Washington Society," *Munsey's Magazine,* March 1908.

O'Taylor, Quaker, "The Loveable Dolley Madison," *National Republic,* April 1928.

Paddock, Charles L., "The White House at Christmas Time," *The Christian Advocate,* December 17, 1931.

Parker, George F., *Recollections of Grover Cleveland.* New York: The Century Co., 1909.

Pendel, Thomas F., *Thirty-six Years in the White House: Lincoln-Roosevelt.* Washington: The Neale Publishing Co., 1902.

Peter, John, "Washington's Official Hosts," *Look,* April 1964.

Phayre, Ignatius, "The White House and Its Occupants," *The Quarterly Review.* London, April 1928.

Poore, Ben: Perley: *Perley's Reminiscences of Sixty Years in the National Metropolis.* Tecumseh, Mich.: A. W. Mills, 1886. Also published in 1886 by Hubbard Bros., Philadelphia, and W. A. Houghton, New York.

Porter, E. R., "Historic Weddings of the White House," *Pearson's Magazine,* March 1906.

Pratt, Harry E., and East, Ernest E., "Mrs. Lincoln Refurbishes The White House," *Lincoln Herald,* February 1945.

Pringle, Henry F., "The President's Musical Marines," *The Saturday Evening Post,* April 14, 1951.

Proxmire, Ellen, *One Foot in Washington.* Washington: Robert B. Luce, Inc., 1964.

Pryor, Mrs. Roger A., *Reminiscences of Peace and War.* New York: The Macmillan Co., 1905.

Quigg, Floyd B., "Preserving White House Traditions," American Forests, June 1949.

Randolph, Mary, *Presidents and First Ladies.* New York: D. Appleton-Century Co., Inc., 1936.

........................, *The Virginia Housewife.* Baltimore: John Plaskitt, 1836.

Remington, Frank L., "The Colorful White House," *American Mercury,* May 1960.

Rinehart, Mary R., "A New First Lady Becomes Hostess for the Nation," *World's Work,* March, 1929.

Robins, Sally Nelson, "President Tyler's Daughter," *The Ladies Home Journal,* March 1894.

Roosevelt, Eleanor, *This I Remember.* New York: Harper & Brothers, 1949.

Ross, Ishbel, *Grace Coolidge and Her Era.* New York: Dodd, Mead & Co. 1962.

......................, *The General's Wife: The Life of Mrs. Ulysses S. Grant.* New York: Dodd Mead & Company, 1959.

Sadler, Christine, *America's First Ladies.* New York: McFadden-Bartell Corporation. 1963.

Sandburg, Carl, *Abraham Lincoln, the War Years.* New York: Harcourt, Brace & Co., 1939.

......................, *Mary Lincoln, Wife and Widow,* Part I by Carl Sandburg; Part II, letters, documents and appendix by Paul M. Angle. New York: Harcourt, Brace & Co., 1932.

Seager, Robert II, *And Tyler, too.* New York, Toronto & London: McGraw-Hill Book Co., Inc., 1963.

Seaton, Josephine, *William Winston Seaton.* Boston: J. R. Osgood & Co., 1871.

Selden, Charles A., "Six White House Wives and Widows," *The Ladies Home Journal,* June 1927.

Shalett, Sidney M., "Dinner With the President," *American Magazine,* February 1948.

Shelton, Isabelle, *The White House Today and Yesterday.* Fawcett Publications, Inc., 1962.

Sinclair, David F., "Monarchial Manners of the White House," *Harper's Weekly,* June 13, 1908.

Singleton, Esther, *The Story of the White House.* New York: The McClure Co., 1907.

Smith, Mrs. Harrison, "The President's House Forty Years Ago," *Godey's Lady Book,* November 1843.

Smith, Helen Barrett, "The First Lady of The White House," *Pearson's Magazine,* March 1909.

Smith, Ira R.T., "My Fifty Years in The White House," *The Post,* December 1948.

Smith, Margaret Bayard, *The First Forty Years of Washington Society, edited by Gaillard Hunt.* New York: Charles Scribner's, Sons, 1906.

Smith, Marie, *The President's Lady: An Intimate Biography of Mrs. Lyndon B. Johnson.* New York, Random House. 1964.

Smith, Marie, and Durbin, Louise, *White House Brides.* Washington, D. C.: Acropolis Books. 1966.

Smith, Theodore C., *The Life and Letters of James Abram Garfield.* New Haven: Yale University Press, 1925.

Starling, Edmund W., *Starling of the White House* as told to Thomas Sugrue. New York: Simon and Schuster, 1946.

Steeholm, Clara and Hardy, *The House at Hyde Park.* New York: The Viking Press. 1950.

Sterling, Peter Roman, "Society in Jefferson's Day," *National Republic,* October, 1929.

Stoddard, William O., *Inside the White House in War Times.* New York: C. L. Webster & Co., 1890.

Strother, French, "A Week in the White House With President Coolidge," *The World's Work,* April 1924.

Sweetser, Kate Dickinson, *Famous Girls of the White House.* New York: Thomas Y. Crowell Company, 1930.

Taft, Helen Herron, "Parties We Gave at The White House," *Delineator,* October, 1914.

......................, *Recollections of Full Years.* New York: Dodd, Mead & Co., 1914.

......................, "White House Debut and a Silver Wedding," *Delineator,* November 1914.

Taylor, A. A. E., "An Afternoon at The White House," *Hours at Home,* December, 1866.

Thayer, William Roscoe, "With Lincoln at The White House," *Harper's Magazine,* January 1915.

The Congressional Club, *Cook Book,* Washington, D. C., 1955.

The Congressional Club, *The Congressional Club Cook Book: Favorite National and International Recipes,* Washington, D. C., 1961; 1965.

Thomas, Kathrine, "How Washington Entertains—Theodore Roosevelt, President," *The Saturday Evening Post,* October 22, 1904.

Thomas, Katherine Elwes, "Two Famous White House Recipes," *Good Housekeeping Magazine,* July 1911.

Thurston, Mynna, "The History of Dolly Payne Madison, Wife of James Madison," December, 1930, Shepherdstown, West Virginia.

Todd, Helen, *A Man Named Grant.* Boston: Houghton Mifflin Co., 1940.

Townsend, George Alfred ("Garth"), *Washington Outside and Inside.* Hartford, Conn., and Chicago: J. Betts & Co., 1873.

Tully, Grace, *F. D. R., My Boss.* New York: Charles Scribner's, Sons, 1949.

Tully, Grace, "Stories and Anecdotes About FDR, My Boss," *The Ladies Home Journal,* July 1949.

Tschirky, Oscar, *The Cookbook by "Oscar" of The Waldorf.* New York: The Saalfield Publishing C., 1908.

Tyler, Lyon G., *The Letters and Times of the Tylers.* Richmond, Va.: Whittet & Shepperson, 1884-1896.

Tyree, Marion Cabell, *Housekeeping in Old Virginia.* Louisville, Ky.: John P. Morton and Company, 1890.

Upton, Harriet Taylor, *Our Early Presidents.* Boston: D. Lothrop Company, 1890.

Wade, Margaret, "White House Parties," *The Delineator,* March 1922.

Ward-Hanna, Elaine, "Perfect Little Dinners From The White House," *The Ladies Home Journal,* November 1962.

Waterhouse, Dorothy S., "Presidential China," *Antiques,* November, 1948.

Webb, Terrell, ed., *Washington Wife, Journal of Ellen Maury Slayden from 1897-1919.* New York and Evanston: Harper & Row, 1962.

Weisinger, Mor, "First Florist of The Land," *The Saturday Evening Post,* September 10, 1949.

Wellman, Rita, "The Capital Makes Merry," *House and Garden,* July 1940.

West, A. G., "Flowers of The White House," *American Motorist,* June 1931.

West, Marian, "Our Four Year Queens," *Munsey's Magazine,* September 1901.

W. G., "The White House as a Social Centre," *Harper's Bazaar,* February 1908.

Wharton, Anne Hollingsworth, *Social Life in the Early Republic.* Philadelphia and London: J. B. Lippincott Co., 1948.

......................., *Salons, Colonial and Republican.* Philadelphia and London: J. B. Lippincott and Co., 1900.

Whitney, Janet, *Abigail Adams.* Boston: Little, Brown and Company. 1948.

Whitton, Mary Ormsbee, *First First Ladies, 1789-1865.* New York: Hastings House 1948.

Wilcox, Mary R., "The White House Chatelaines," *D.A.R. Magazine,* February 1916.

Wilcox, Uthai Vincent, "Tales of the White House," *The Farm Journal,* December 1925.

......................., "The White House Remembers," *The St. Nicholas Magazine,* March 1929.

Willets, Gilson, *Inside Story of the White House.* New York: The Christian Herald, 1908.

Williams, Charles R., *The Life of Rutherford Birchard Hayes.* Boston and New York: Houghton Mifflin Co., 1914.

Wilson, Edith Bolling, *My Memoir.* Indianapolis: The Bobbs-Merrill Company 1939.

Wilson, Flora, "Diplomats and Dinners in The White House," *Success Magazine,* August 1927.

......................., "The Social Jealousy in and About the White House," *Success Magazine,* July 1927.

Women's National Press Club, *Who Says We Can't Cook!* Washington: McIver Art and Publications, Inc., 1955.

................................., *Second Helping.* Washington, D. C.: McIver Art and Publications, Inc., 1962.

Woodruff, Caroline S., "A Visit to the White House," *N.E.A. Journal,* May 1928.

Ziemann, Hugo and Mrs. F. L. Gillette, *The White House Cookbook.* Akron, Ohio: The Saalfield Publishing Company, 1925.

* "A Glimpse at Life in Today's White House," *U. S. News and World Report,* April 3, 1961.

* "A Reception at The White House," *The New York Times Magazine,* May 13, 1962.

* "Biggest Reception Ever," *Ebony,* May, 1963.

* "Domestic Science in The White House," *Harper's Bazaar,* February 16, 1901.

* "Entertaining People: First Hospitality Lady," *McCall's,* November 1966.

* "First Ladies of the Land, Past and Present," *The Literary Digest,* August 15, 1925.

* "Glamour Era in The White House," *U. S. News and World Report,* May 14, 1962.

* "Ike and Mamie in The White House," *U. S. News and World Report,* January 23, 1953.
* "Memories," *The New Yorker,* January 27, 1962.
* "More Merry-Go-Round: Another President's Daughter," *Hearst's International Cosmopolitan,* February 1932.
* "New Year's Day at The White House," *The Mail and Express Illustrated Saturday Magazine,* January 9, 1904.
* "Open House," *The Literary Digest,* July 3, 1937.
* "President's Functions," *Harper's Weekly,* March 24, 1894.
* "Sixteen Hundred Pennsylvania Avenue; What Goes On at The White House," *Red Book,* February 1937.
* "Social Life in Washington During Mr. Madison's Presidency," *Every Saturday,* April 1, 1871.
* "Statesmen's Dishes and How to Cook Them." Washington, D. C.: *National Tribune,* 1890.
* "The Battle of Bladensburg and Dolly Madison," *American Monthly Magazine,* November 1892.
* "The Daughter of Andrew Johnson," *The Ladies Home Journal,* March 1894.
* "The First Mistress of The White House," *National Republic,* November 1926.
* "The Last Presidential Drawing Room," *Harper's Weekly,* May 8, 1858.
* "The White House as a Social Treadmill," *The Literary Digest,* August 20, 1927.
* "The Worries of the First Mistress of the White House," *National Republic,* June 1926.
* "There's a New Social Style at The White House," *U. S. News and World Report,* February 17, 1964.
* "Washington Social Life During the Times of Madison and Monroe," *Every Saturday,* April 1, 1871.
* "White Tie Winter," *New Republic,* December 9, 1946.

* Author not listed.

Files of the following periodicals and newspapers contemporary with the various presidential administrations:

HARPER'S WEEKLY
LESLIE'S ILLUSTRATED WEEKLY
NATIONAL INTELLIGENCER
NEW YORK DAILY GRAPHIC
NEW YORK HERALD
NEW YORK TIMES
NEW YORK WORLD
PHILADELPHIA PRESS
EVENING STAR, Washington, D. C.
EVENING SUN, Baltimore, Md.
WASHINGTON DAILY NEWS
WASHINGTON POST
WASHINGTON TIMES-HERALD

Acknowledgements

My sincere gratitude goes to many people who contributed in various ways to the preparation of this book. But my deepest thanks go to the Presidents and First Ladies whose hospitality in The White House made it possible.

I wish to especially acknowledge with thanks also the cooperation of James Ketcham, White House curator; Elizabeth Carpenter, press secretary and staff director to Mrs. Lyndon B. Johnson; Bess Abell, Social Secretary, and other members of the White House staff; Evelyn Lincoln, secretary to the late President John F. Kennedy; Pamela Turnure Timmons, press secretary to Mrs. John F. Kennedy; Pierrette Spiegler, also on Mrs. Kennedy's White House staff; Mrs. Ralph H. Dillon, director of the Woodrow Wilson House; Margaret Klapthor, associate curator, Division of Political History, The Smithsonian Institution; the staffs of the Library of Congress, the National Archives, and the District of Columbia Public Library; Elizabeth Drewry, director, Franklin D. Roosevelt Memorial Library; Mrs. Harold Kraft, museum curator, Sagamore Hill National Historic Site; Mrs. Gordon Parker, curator, Wheatland, the home of President James Buchanan; Mrs. T. P. Yeatman, Hamilton Place, Columbia, Tennessee; Watt P. Marchman, director, The Rutherford B. Hayes Library; Nellie Jane Macdonald; and members of the staff of other libraries and memorials to former presidents who provided information contained herein.

My thanks, also, to The Women's National Press Club, Eve Edstrom, president, for permission to reprint from its cookbook *Second Helping* the following recipes: Mrs. Dwight D. Eisenhower's Danish Tomatoes and Mrs. John F. Kennedy's Bombe Glace Mandarin; and from its cookbook, *Who Says We Can't Cook!* Mrs. Harry S. Truman's Frozen Lemon Pie, Mrs. Dwight D. Eisenhower's Million Dollar Fudge, and Mrs. Calvin Coolidge's Spinach Timbales.

To The Congressional Club, Mrs. Roy A. Taylor, president, for permission to reprint from *The Congressional Club Cookbook* the following recipes: Jacqueline Kennedy's Fish Chowder (1961 edition); Bess Truman's Cheese Ring (1961 edition); Eleanor Roosevelt's Toasted Cheese Sandwich and Blueberry Pudding (1961 edition) and Mamie D. Eisenhower's Cold Curry Soup (1965 edition).

For inspiration and encouragement, I am indebted to my mother, Dessie Marr Smith, to whom this work is lovingly dedicated.

M. S.

PHOTO CREDITS

Copyright by White House Historical Association. Photographs by National Geographic Society—Pages: Title Page, 2-3, 9, 10, 13, 15, 16, 215

White House Collection. Copyright by The White House Historical Association—Pages: 54, 70, 75, 141

Copyright by White House Historical Association— Pages: 6, 17, 50, 138, 179, 206, 207, 226, 235

District of Columbia Library—Page: 138

Library of Congress—Pages: 29, 47, 52, 63, 82, 86, 90, 94, 96, 98, 100, 108, 110, 112, 115, 118, 119, 126, 130, 131, 135, 137, 145, 147, 148, 151, 152, 158, 161, 177, 191, 228

Perley's Reminiscences—Pages: 25, 57, 71, 81, 84, 91, 101

Franklin D. Roosevelt Memorial Library—Pages: 198, 202

The Smithsonian Institution—Page: 25

Roy Teller—Pages: 162, 301

The Washington Post—By Joe Heiberger, Staff Photographer—Pages: 172, 194, 216, 243, 253, 272

INDEX

President	Term	Hostesses		Salary	Outstanding Social Innovations
George Washington	4/30/1789—3/3/1797	Martha Washington	NO PERSONAL, TRAVEL OR OTHER EXPENSE ALLOWANCES	$25,000	Gave weekly levees or drawing rooms. Bowed in salute to his guests.
John Adams	3/4/1797—3/3/1801	Abigail Adams		$25,000	
Thomas Jefferson	3/4/1801—3/3/1809	Martha Randolph Mary Eppes Dolley Madison		$25,000	Dispensed with weekly drawing rooms in favor of two public receptions a year, on New Year's and July 4th. Shook hands with his guests, instead of bowing. Used a "dumb waiter" for serving, his own invention.
James Madison	3/4/1809—3/3/1817	Dolley Madison		$25,000	
James Monroe	3/4/1817—3/3/1825	Elizabeth Monroe		$25,000	First Lady did not make first calls or return calls.
John Quincy Adams	3/4/1825—3/3/1829	Louisa Adams		$25,000	Encouraged dancing. Louisa passed from group to group to welcome her guests, rather than in a fixed line surrounded by cabinet wives.
Andrew Jackson	3/4/1829—3/3/1837	Emily Donelson		$25,000	
Martin Van Buren	3/4/1837—3/3/1841	Angelica Van Buren		$25,000	Dispensed with public receptions except for New Year's Day.
William Henry Harrison	3/3/1841—4/4/1841	Jane Harrison Mrs. James Findlay		$25,000	
John Tyler	4/6/1841—7/9/1845	Priscilla Tyler Julia G. Tyler		$25,000	Dancing of the waltz and polka. Initiated the playing of "Hail to the Chief" by the Marine Band.
James K. Polk	3/4/1845—3/3/1849	Sarah Polk		$25,000	Gaslights installed in the White House.
Zachary Taylor	3/5/1849—7/9/1850	Betty Bliss		$25,000	
Millard Fillmore	7/10/1850—3/3/1853	Abigail Fillmore		$25,000	
Franklin Pierce	3/4/1853—3/3/1857	Mrs. Abby Kent Means		$25,000	
James Buchanan	3/4/1857—3/3/1861	Harriet Lane		$25,000	Conservatory built.
Abraham Lincoln	3/4/1861—4/15/1865	Mary Todd Lincoln		$25,000	State dinners prepared by the White House staff rather than caterers.
Andrew Johnson	4/15/1865—3/3/1869	Martha Patterson Mary Stover		$25,000	Abandoned the practice of announcing names as guests passed through the receiving line.
Ulysses S. Grant	3/4/1869—3/3/1877	Julia Dent Grant		$50,000	
Rutherford B. Hayes	3/3/1877—3/3/1881	Lucy Hayes		$50,000	Returned to the practice of having cabinet wives assist the President's wife in receiving. Originated the Easter Monday egg roll.
James A. Garfield	3/4/1881—9/19/1881	Lucretia Garfield		$50,000	
Chester A. Arthur	9/20/1881—3/3/1885	Mrs. John McElory		$50,000	Dinner hour changed from seven to eight. Courses reduced to fourteen. Recognized achievers by inviting them to dinner at the White House.